Modern Physiology and Anatomy for Nurses

Modern Physiology and Anatomy for Nurses

JOHN GIBSON MD

SECOND EDITION

Blackwell Scientific Publications
OXFORD LONDON EDINBURGH
BOSTON MELBOURNE

© 1975, 1981 by
Blackwell Scientific Publications
Editorial offices:
Osney Mead, Oxford OX2 0EL
8 John Street, London WC1N 2ES
9 Forrest Road, Edinburgh EH1 2QH
52 Beacon Street, Boston
  Massachusetts 02108, USA
214 Berkeley Street, Carlton
  Victoria 3053, Australia

First published 1975
Second edition 1981

Printed and bound
in Great Britain by
Butler & Tanner Ltd
Frome and London

DISTRIBUTORS

USA
  Blackwell Mosby Book Distributors
  11830 Westline Industrial Drive
  St Louis, Missouri 63141

Canada
  Blackwell Mosby Book Distributors
  120 Melford Drive, Scarborough
  Ontario M1B 2X4

Australia
  Blackwell Scientific Book Distributors
  214 Berkeley Street, Carlton
  Victoria 3053

British Library
Cataloguing in Publication Data
Gibson, John
  Modern physiology and anatomy for
    nurses.—2nd ed.
  1. Physiology 2. Anatomy, Human
  I. Title
  612′0024613   QP34

ISBN 0–632–00795–8

Contents

Preface

For the second edition this book has been revised and brought up to date throughout. Much of the information has been tabulated in order to facilitate learning.

J. G.

1
The Body and its Cells

Physiology is the study of the functions of the body.
Anatomy is the study of the form of the body.
Histology is the study of micro-anatomy, i.e. the study of the cells of the body with the aid of a microscope.
Biochemistry is the study of the chemistry of living structures.
Embryology is the study of development before birth.

THE CELL

The *cell* (Fig. 1.1) is the structural unit of which tissues are composed. Cells vary in shape, size and content, according to their varying functions. Basically all cells are composed of:

A *cell membrane*: a thin, flexible membrane enclosing the rest of the cell, and composed of protein and lipid molecules.

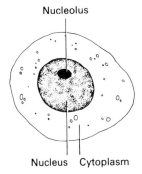

Nucleolus

Nucleus   Cytoplasm

Fig. 1.1. The basic form of a cell, showing cytoplasm enclosed within a membrane, nucleus and nucleolus.

Cytoplasm: the substance of which most of the cells is usually composed. It may contain granules and a fine network.

A *nucleus*: a special rounded dense structure within the cytoplasm. It is surrounded by a membrane and contains particles of gene-carrying chromatin.

The *nucleolus* is a small round body within the nucleus. It is composed mainly of ribonucleoprotein.

The mature red blood cell has no nucleus, its nucleus having been extruded from it in the course of its development. Blood platelets are small particles which have no nucleus.

The function of cells varies according to their basic purpose, e.g. the cells of the liver are engaged in producing chemical changes, the cells of the heart muscle are engaged in contracting and relaxing, the cells of the thyroid gland are engaged in producing hormones. Cells do not usually touch each other, but are separated by a thin, fluid-occupied space. Water and chemical substances pass from fluid to cell and from cell to fluid, the cell membrane exercising a selective permeability, letting some substances through and not others.

Essential features of function in any cell are:

(a) Cell activity requires oxygen and produces carbon dioxide.

(b) The ability to generate chemical energy from foodstuffs, to take out of the fluid around it what it requires, and to maintain and repair itself.

(c) Enzymes (catalysts which speed up chemical reactions without themselves being changed) are present in cells as proteins. Each enzyme has a specific function; it promotes a specific chemical reaction.

(d) DNA (deoxyribonucleic acid) and RNA (ribonucleic acid) are necessary for the reproduction of a cell.

THE TISSUES

> *Main types of tissue*
>   epithelium
>   connective tissue
>   muscle
> blood cells
> nervous tissue

Epithelium

Epithelium covers the external and internal surfaces of the body and lines the glands and ducts opening on to these surfaces.

The cells of which epithelium is composed are:
squamous (flat),
cuboidal (cube-shaped) (Fig. 1.2),
columnar (tall).

Simple squamous epithelium (endothelium) is composed of a single layer of flat cells. These cells line the inside of the heart, blood vessels and lymph vessels, and the alveoli (air sacs) of the lungs.

Stratified epithelium (Fig. 1.3) is composed of more than one layer of squamous cells, e.g. the skin, where the top layers are being constantly worn away.

Transitional epithelium is a form of epithelium about four cells thick found in the collecting tubules of the kidneys, the ureters, bladder and part of the female urethra.

Cuboidal epithelium (Fig. 1.2) forms the walls of the follicles in the thyroid gland, from which cells it produces its hormones.

Columnar epithelium lines the inner surface of the stomach and intestines, forming their mucous membrane.

Ciliated columnar epithelium (Fig. 1.2) has cilia protruding from its surface. *Cilia* are thin movable processes which are engaged in sweeping movements, e.g. in keeping dust out of the lungs and moving an ovum along a uterine tube.

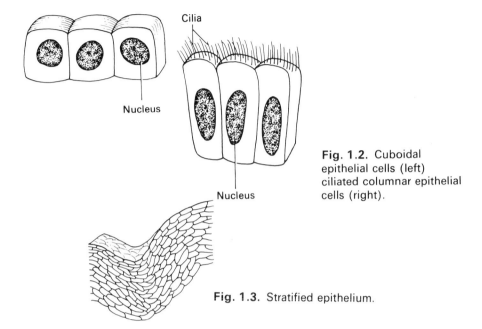

Cilia

Nucleus

Nucleus

Fig. 1.2. Cuboidal epithelial cells (left) ciliated columnar epithelial cells (right).

Fig. 1.3. Stratified epithelium.

STRUCTURES

The *parenchyma* of an organ is the essential functioning tissue of which the organ is composed (contrasted with supporting tissue, blood vessels etc) (Fig. 1.4).

A *mucous membrane* is epithelium moistened by mucin, a sticky substance produced by glands. Mucous membranes line:

the inside of the nose and nasal sinuses,

the trachea and bronchi,

the mouth,

the oesophagus, stomach and intestine.

A *serous membrane* is a thin membrane of simple squamous epithelium which lines three cavities in the trunk and encloses many of the organs within them. The serous membranes—and their contents—are:

the *pericardium*—the heart,

the *pleura*—a lung,

the *peritoneum*—many of the abdominal organs.

A *basement membrane* is a layer of condensed substance present between epithelium and connective tissue.

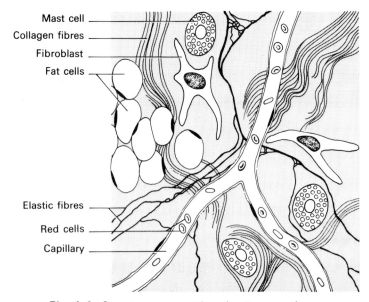

Mast cell
Collagen fibres
Fibroblast
Fat cells

Elastic fibres
Red cells
Capillary

Fig. 1.4. Structures present in subcutaneous tissue.

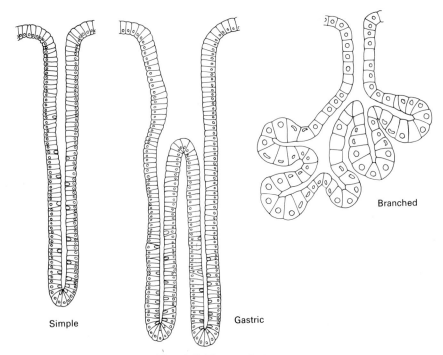

Simple   Gastric

Branched

Fig. 1.5. Types of glands.

GLANDS

Glands (Fig. 1.5) are of two types:

endocrine glands: from which hormones, their secretions, pass into the blood,

exocrine glands: from which secretions pass to an external or internal surface of the body, either directly or through a duct. They vary in complexity from simple tubes to compound glands of many tubes.

Some glands (e.g. the pancreas) are of both types. The pancreas produces two hormones (insulin and glucagon) which pass into the blood and pancreatic juice which passes along a tube into the duodenum.

Connective tissue

The essential functions of connective tissue are binding together, supporting and protecting. Connective tissue can be soft or dense.

Connective tissue
fibrous tissue
areolar tissue
fat
tendon
cartilage
bone

SOFT CONNECTIVE TISSUE

Soft connective tissue (Fig. 1.4) is made up of:

ground substance: an amorphous material enclosing various cells, fibres, capillaries, etc.

cells: these can be:

(a) *fibroblasts*: cells from which fibres develop,

(b) *plasma cells*: large oval cells,

(c) *histiocytes*: scavenger cells, capable of absorbing micro-organisms and small particles,

(d) *fat cells*: large cells with much fat in the cytoplasm,

(e) *mast cells*: large cells with large granules in the cytoplasm.

fibres: these can be:

(a) *elastic fibres*: long, branching fibres,

(b) *collagen fibres*: long, unbranching fibres.

Fibrous tissue: formed mostly of fibres and found in:

ligaments around joints,

sheaths of muscles,

fascia, which divide a structure into various compartments,

outer covering of organs,
dura mater, one of the coverings of the brain.
Areolar tissue: a loose tissue with relatively few fibres found:
 surrounding blood vessels and nerves,
 under the skin,
 under mucous membranes.

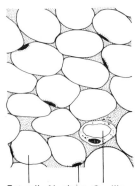

Fig. 1.6. Fatty tissue composed mostly of fat cells.

Fat cell Nucleus Capillary

Fat: contains a large number of fat cells (Fig. 1.6, is the reservoir of fat in the body and is found:
 under the skin,
 between layers of peritoneum in the abdomen,
 at the back of the abdominal cavity.

DENSE CONNECTIVE TISSUE

Forms:
 tendons: straps or cords attached at one end to a muscle and at the other to bone and composed of fibres packed closely together and lying in the long axis of the tendon,
 cartilage (see pp. 12–13),
 bone (see pp. 9–11).

Muscle

Muscle cells have the ability to contract and relax. There are three types:
 skeletal muscle (see pp. 65–8),
 smooth muscle (see p. 65),
 cardiac muscle (see p. 87).

Blood cells

Blood cells are of three main types: red cells, white cells and platelets.

Nervous tissue

Nervous tissue is composed of nerve cells and their attached fibres. It is found in the brain, spinal cord, peripheral nerves and autonomic nervous system.

THE ANATOMICAL POSITION

In order that the relationship of one structure to another can be described consistently, it is necessary to establish a fixed anatomical position. Whatever the actual position of the body or any part of it, the relationship of one structure to another is described as it is in the anatomical position.

The *anatomical position* is one in which

(a) the person is erect and looking forwards,
(b) the arms are by the sides with the palms of the hands turned forwards,
(c) the legs and feet are together with the feet pointing forwards.

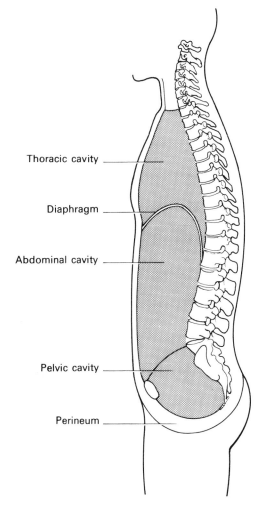

Thoracic cavity

Diaphragm

Abdominal cavity

Pelvic cavity

Perineum

Fig. 1.7. The cavities of the body.

Terms

Anterior, posterior, superior and *inferior* describe a structure as being in front, behind, above or below another structure.

Medial and *lateral* mean nearer to or further from a vertical midline drawn through the centre of the head and trunk and between the legs and feet.

Internal and external mean within or outside another structure. These terms are sometimes used instead of medial and lateral.

Proximal and *distal* mean nearer to or further from the origin or centre of a structure.

Palmar refers to the palm of the hand. *Plantar* refers to the sole of the foot. The *dorsum* of a foot is its upper surface.

2
Bones and Cartilage

The skeleton is a jointed structure of bones and cartilage.

Functions of the skeleton
provision of levers for muscles to act on
protection of organs (skull protects the brain;
thorax protects the heart, lungs and great
vessels; pelvis protects the pelvic organs)
formation of red blood cells in bone marrow
store of calcium and phosphorus, drawn on
when required

BONE

General features
Bones are composed of:

compact bone (Fig. 2.1), a dense hard outer layer, enclosing the rest of the bone. The bone is arranged around long canals called haversian canals.

cancellous (spongy) bone, a honeycomb of bone inside the compact bone, with bars of bone arranged in patterns capable of bearing the weight and stresses to which any particular bone is subjected.

a canal, in long bones, a tube in which there is no bone.

Cancellous bone and the canals are filled with bone marrow—red bone marrow, which is blood-forming tissue, or yellow bone marrow, which is fat. Red bone marrow is present in all bones in the fetus and young child; but in adult life it is replaced by yellow bone marrow except at the upper ends of humerus and femur, the vault of the skull, the vertebrae, the ribs, the sternum and the pelvic bones, in which red bone marrow persists.

Bone is composed of cells and a matrix.

Bone cells are either *osteoblasts*, which when mature are called osteocytes and form bone, or *osteoclasts*, which destroy bone. By the actions of these cells bones are in a continuous state of formation and destruction.

The *matrix* is composed of ground substance, fibres and salts. The proportion of salts to organic matter is:

calcium phosphate and other salts—70 per cent,
organic matter—30 per cent.

9

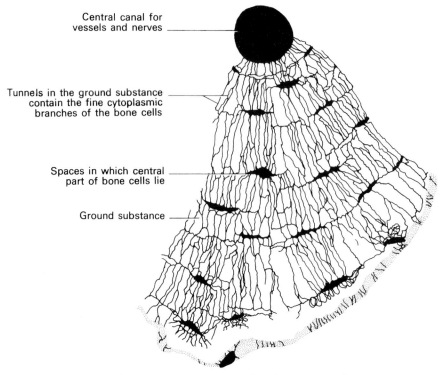

Central canal for
vessels and nerves

Tunnels in the ground substance
contain the fine cytoplasmic
branches of the bone cells

Spaces in which central
part of bone cells lie

Ground substance

Fig. 2.1. The haversian system forming compact bone.

The *periosteum* is a tough and vascular membrane of connective tissue, closely attached to the outside of bones, except within the joints, where the bone is covered with cartilage. Its innermost cells manufacture bone cells. The periosteum is essential for thickening bone and the repair of fractures.

Blood supply

Bone requires a good blood supply (Fig. 2.2), for its own nutrition and for the manufacture, in certain bones, of red blood cells. Arteries enter bone at:
(a) the nutrient canal, a hole in the shaft of long bones, usually about half way along.
(b) through several openings situated close to the ends of bones.

Fractures

Fractures can be:
   simple: the bone is broken in one place; other structures are not involved,
   comminuted: the bone is broken into several pieces,
   compound: a fractured end of bone is exposed to air, e.g. by sticking through the skin,

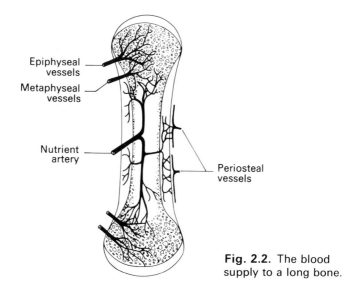

Epiphyseal vessels

Metaphyseal vessels

Nutrient artery

Periosteal vessels

Fig. 2.2. The blood supply to a long bone.

impacted: the broken ends are driven into each other,

greenstick: a partial fracture occurring in childhood when the bones are incompletely ossified,

fracture-separation of an epiphysis: occurs in children, an epiphysis (usually with a bit of the shaft attached to it) is torn off.

The *treatment* of a fracture is by *reduction*—the pulling or pushing of the broken pieces into their correct alignment and by a period of *immobilization* by splinting.

The *healing* of a fracture is achieved by:

(a) the formation of a blood clot between the broken pieces:

(b) fibrous tissue developing in the clot:

(c) osteoblasts from the broken ends and the periosteum invading the fibrous tissue and forming callus, new bone:

(d) the callus, which is at first softer and larger than the normal bone, being converted into normal bone and excess callus being absorbed; the bone is thus remoulded to its original shape and size.

This healing process is achieved by a protein (BMP) in bone, which, when the bone is fractured, diffuses out into the surrounding tissues and converts connective tissue cells into bone cells. Bone is the only tissue which can heal itself without forming a scar. In old people there is less **BMP** and fractures heal more slowly.

Abnormalities of development

Achondroplasia is a condition in which the epiphyses fail to develop. Affected persons have a head and trunk of normal size, but very short limbs.

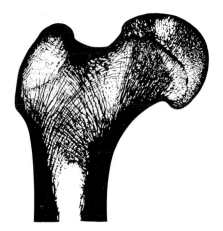

Fig. 2.3. Section through the upper end of the femur, showing the struts of bone within the head, neck and great trochanter, and the upper end of the medullary cavity.

Diseases

Osteomyelitis is inflammation of bone. *Periostitis* is inflammation of the periosteum.

CARTILAGE

Cartilage (Fig. 2.4) is composed of cells and fibres enclosed within a solid matrix or substance. It is tough but has a greater degree of elasticity and compressibility than bone. It has no blood vessels, lymph vessels or nerves. There are three kinds of cartilage, with differing cell and fibre content.

Hyaline cartilage

Hyaline cartilage forms:
(a) the cartilage in which most bones are formed,
(b) the costal cartilages, at the anterior end of the ribs; in later life this cartilage ossifies (i.e. becomes bone),
(c) articular cartilage within the joints. This cartilage covers the joint surface of bones and is smooth, pearly grey and glistening with fluid secreted by the synovial membrane of the joint,
(d) the larynx and rings of the trachea. The cartilage of the larynx can become ossified.

Elastic cartilage

This is similar to hyaline cartilage but contains many elastic fibres. It is found in: the pinna of the ear, the epiglottis, and the auditory (pharyngo-tympanic) tube.

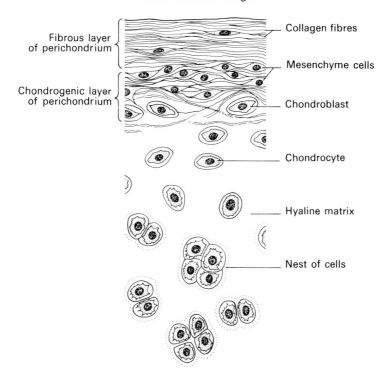

Fibrous layer of perichondrium

Chondrogenic layer of perichondrium

Collagen fibres

Mesenchyme cells

Chondroblast

Chondrocyte

Hyaline matrix

Nest of cells

Fig. 2.4. Cartilage with its outer layer of perichondrium.

Fibrocartilage

This contains a great many collagen fibres, threads found in connective tissue. Plates of it, acting like buffers between adjacent bony surfaces, are found in:
(a) the intervertebral discs, the outer ring of which is formed of fibrocartilage,
(b) the pubic symphysis, separating the two pubic bones at this point,
(c) the semilunar cartilages of the knee joint,
(d) the discs of the temporomandibular and sternoclavicular joints.

3
The Skull

The skull is formed by the union of several bones. Individual bones (except the mandible) are united at sutures. The *sutures* are formed of a thin layer of fibrous tissue into which the serrated edges of bones are locked. They become ossified from 35 years onwards. In the roof of the skull the outer and inner surfaces are formed of compact bone with a spongy layer called the diploe between them. There is a marked variation of thickness of bone in the skull in different individuals. The skull is thickest where it is not protected by muscle.

Functions of skull
The protection of the brain and of the special senses of sight and hearing
provision of attachment for muscles acting on the head
provision of a mounting for the teeth

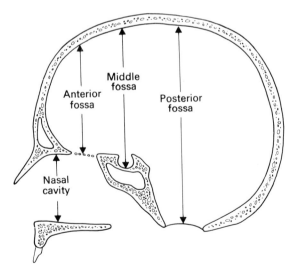

Fig. 3.1. The fossae of the skull.

> *Bones of skull*
>   frontal
>   parietal, right and left
>   occipital
>   temporal, right and left
>   ethmoid
>   sphenoid
>   maxilla
>   mandible
>   zygomatic, right and left
>   palatine, right and left
>   nasal, right and left
>   lacrimal, right and left
>   vomer
>   conchae, right and left

Wormian bones are present in some skulls: they are small bones found in the suture-lines and are a result of development from separate centres of ossification.

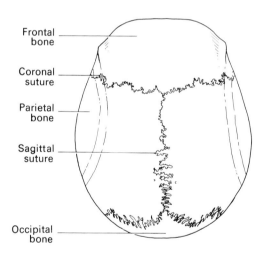

Frontal bone
Coronal suture
Parietal bone
Sagittal suture
Occipital bone

Fig. 3.2. the skull from above.

The skull viewed from above (Fig. 3.2) *shows*:
frontal bone in front,
right and left parietal bones,
occipital bone behind.

The skull viewed from behind (Fig. 3.3) *shows*:
right and left parietal bones,
occipital bone at the back,
mastoid processes of the temporal bones.

Chapter 3

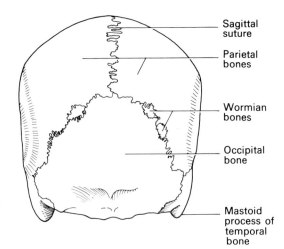

Fig. 3.3. The skull from behind. Wormian bones are variable in size and number.

The skull viewed from the side (Fig. 3.4) *shows*:
the vault—formed of frontal, parietal, temporal and occipital bones,
the face—formed of frontal, nasal, zygomatic, temporal bones, maxilla and mandible.
    The skull viewed from the front (Fig. 3.5) *shows*:
the vault—formed of the frontal bone,
the face—formed of the frontal, nasal, zygomatic bones, maxilla and mandible.

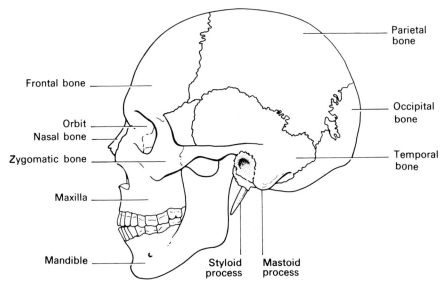

Fig. 3.4. The skull from the side.

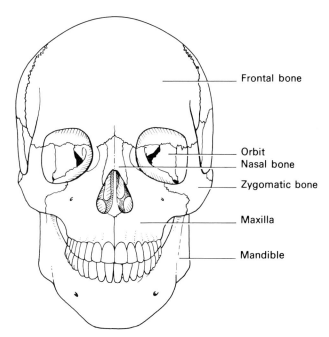

Fig. 3.5. The skull from the front.

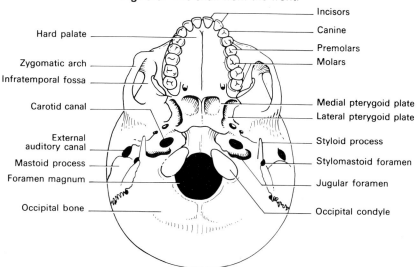

Fig. 3.6. The skull from below.

The skull viewed from below (*with the mandible removed*) (Fig. 3.6) *shows*:
(a) the upper dentition, mounted on the hard palate, which is composed of the palatal processes of the maxilla in front and part of the palatine bones behind,
(b) the zygomatic arch, composed mainly of the zygomatic bone, with a projection from the maxilla in front and another from the temporal bone behind,

(c) the infratemporal fossa, the space between the zygomatic arch and the vault of the skull,

(d) part of the sphenoid bone,

(e) part of the temporal bone, including the mastoid process, the styloid process and the carotid canal (through which passes the internal carotid artery),

(f) the occipital bone, surrounding the foramen magnum,

(g) the foramen magnum, through which pass the spinal cord and vertebral arteries.

(h) the occipital condyles, on either side of the foramen magnum, for articulation with the atlas (1st cervical vertebra),

(i) several foramina for vessels and nerves.

The skull viewed from within, with the vault removed, shows: three terraces on either side:

the anterior cranial fossa,

the middle cranial fossa,

the inferior cranial fossa.

Anterior cranial fossa is composed of:

(a) the orbital plates of the frontal bone, which form the roofs of the orbit,

(b) the crista galli and cribriform plate of the ethmoid bone, lying between the two orbital plates,

(c) the lesser wing of the sphenoid bone.

The optic foramen is the hole in the sphenoid bone through which the optic nerve of each side passes from the orbit towards the brain.

Middle cranial fossa is composed of:

(a) the sphenoid bone in front,

(b) the temporal bone in the middle,

(c) the front of the petrous part of the temporal bone behind.

The *pituitary fossa* is a central depression in the sphenoid bone which in life encloses the pituitary gland. The anterior and posterior clinoid processes are tiny projections of bone, two in front of and two behind the pituitary fossa.

The internal carotid artery enters the skull through an opening posterolateral to the fossa.

Posterior cranial fossa, the largest and deepest, is composed of:

(a) the back of the petrous part of the temporal bone,

(b) the occipital bone.

It shows:

(a) a deep depression on either side in which lies the cerebellum,

(b) the foramen magnum,

(c) grooves in which lie venous sinuses.

Sutures

Some sutures are named:

the *coronal suture* separates the frontal bone from the parietal bones,

the *sagittal suture* separates the two parietal bones,

the *lambdoid suture* separates the two parietal bones from the occipital bone.

Clinical features

The bony vault of the skull has some degree of elasticity, and may give a little under a blow which can damage the brain without necessarily fracturing the skull.

A fracture of the skull is usually the result of a direct blow. A localized blow on the vault is likely to cause a fracture in which an area of bone is driven inwards; the brain, its blood vessels and the meninges around the brain are likely to be damaged.

The base of the skull is rigid and a fracture through it is likely to injure other structures—the air sinuses, the blood vessels which pass through it, and the cranial nerves.

Frontal bone

The frontal bone forms: (a) the front of the vault, (b) most of the roof of the orbits (the bony cavity in which lies the eye), and (c) part of the inner wall of the nasal cavity.

The part forming the front of the vault extends from the *supra-orbital margins*, the upper borders of the orbits, until it meets the two parietal bones at the coronal suture.

The *frontal sinuses* are two air spaces within the frontal bone just above the orbits. They communicate down a narrow tube with the middle meatus of the nasal cavity on each side, and in life are lined with mucous membrane continuous with the mucous membrane inside the nose. They develop from the first year of life up to puberty; they often vary in size and one may be much larger than the other and diverge towards the opposite side. In this part the frontal bone is thick because it contains these sinuses.

The inner surface of the frontal bone is grooved for blood vessels and is separated by membranes from the frontal lobes of the brain.

Parietal bones

There are two parietal bones, right and left. They are flat curved bones, which form the greater part of the sides and top of the skull and some of the back. They are separated from each other by the sagittal suture, from the frontal bone in front by the coronal suture, from the occipital bone behind by the lambdoid suture, and below by a suture from the temporal bone.

The inner surface is marked by furrows for blood vessels and is separated by membranes from the parietal lobes of the brain.

Occipital bone

The occipital bone forms the back of the vault of the skull and the posterior part of its base.

The *foramen magnum* is a large hole in the occipital bone, through which pass:
(a) the top end of the spinal cord where it joins the medulla oblongata of the brain, and
(b) the right and left vertebral arteries which provide some of the blood for the brain.

The *occipital condyles* are the two joint surfaces, one on each side of the foramen magnum, which articulate with joint surfaces on the upper surface of the atlas (1st cervical vertebra). The hypoglossal (12th cranial) nerve on each side passes through a canal in the base of each condyle.

The inner surface of the bone shows furrows for blood vessels and is separated by membranes from the occipital lobes of the brain and the cerebellum.

Temporal bones

Each temporal bone, right and left, is a complex bone, which forms part of the vault and part of the base of the skull.

Prominent features on the *outer surface* are:
(a) a flat part forming part of the vault and separated by sutures from the frontal bone in front, the parietal bone above, and the occipital bone behind,
(b) the external auditory meatus, an opening leading down within the bone to the middle ear,
(c) the mastoid process, just behind the ear, a peg of bone, variable in size, and containing the mastoid air sinus.
Within the skull the temporal bone shows a prominent rough ridge of bone, running backward and outwards. The organs of hearing and balance are enclosed within this ridge. The facial (7th cranial) nerve and the auditory (8th cranial) nerve enter the ridge at the internal auditory meatus. The facial nerve runs through the temporal bone to emerge on the inferior surface of the skull just medial to the mastoid process. The auditory nerve ends in the organs of hearing and balance.
(d) a thin styloid process, projecting downwards and forwards from the inferior surface; it is an attachment for some small muscles and ligaments.

The *internal carotid artery* on each side passes through the temporal bone to reach the interior of the skull. It enters the carotid canal in the base of the skull and follows an S-shaped course within the bone.

Ethmoid bone

The ethmoid bone, lying between the frontal bone in front and the sphenoid bone behind, forms part of the anterior cranial fossa of the skull, the nasal cavities and the cavity of the orbit. It has
(a) a vertical plate in the midline which forms part of the nasal septum,
(b) a lateral mass on either side, forming part of the outer wall of the nose and the inner wall of the orbit and containing the ethmoid air cells which communi-

cate with the inside of the nasal cavity and in life are lined with mucous membrane.

Within the skull it shows in the midline the crista galli, a small pointed spike of bone, to which is attached the front end of the falx cerebri (see p. 232) and on either side of it a flat plate of bone perforated by several holes through which pass the fibres of the olfactory (1st cranial) nerve from the nose to the brain.

Sphenoid bone

The sphenoid bone lies in the base of the skull and consists mainly of a central block, two wings (greater and lesser) which run outwards from it, and two vertical plates. The central block is called the body of the sphenoid. Part of the greater wing of the sphenoid is visible on the side of the skull where it rises up to meet the parietal bone.

The *body* of the sphenoid lies in the midline of the base of the skull and shows:

the *pituitary fossa*, a deep depression occupied by the pituitary gland,

anterior and *posterior clinoid processes*, two in front and two behind the pituitary fossa,

the *sphenoid air sinus*, variable in size and partly divided into two by a bony septum; it communicates with a space at the back of the nose and in life is lined with mucous membrane continuous with that of the nose.

Clinical features
The pituitary fossa and clinoid processes are plainly visible on a lateral X-ray of the skull. The pituitary gland, which occupies the fossa, can become enlarged, and when that occurs an X-ray shows a 'ballooning' of the fossa and decalcification (disappearance of calcium) in the clinoid processes.

Maxilla

The two maxillae form the upper jaw and all the bone between the eyes and the mouth. They carry the teeth and also form part of the roof of the mouth, the floor of the orbit and the outer wall of the nasal cavity.

The *maxillary antrum* is a large air sinus which occupies most of the body of the maxilla. It communicates with the nasal cavity through an opening high up into the middle meatus of the nose and in life is lined with mucous membrane.

Clinical features
The mucous membrane of the antrum can become infected from the nose. If pus forms, this cannot readily escape because the opening into the nose is so high. It is sometimes necessary to make a hole lower down so that the pus can escape or be washed out; this is done by perforating the thin bone between the antrum and the lower meatus of the nose.

The *zygomatic bones*, right and left, form the prominence of the cheeks and each forms part of an arch, articulating behind with the temporal bone, in front with the frontal bone and below with the maxilla. It may be broken by a severe blow on the cheek.

The *nasal bones*, right and left, form the bridge of the nose, articulating with each other in the middle line.

The *palatine bones*, right and left, are L-shaped and form the back of the hard palate and part of the lateral wall of the nasal cavity.

The *lacrimal bones*, right and left, are small bones in the front of the inner wall of the orbit on each side. The lacrimal fossa is a depression in the bone in which is lodged the lacrimal sac of the eye.

The *vomer* forms most of the bony part of the nasal septum.

The mandible

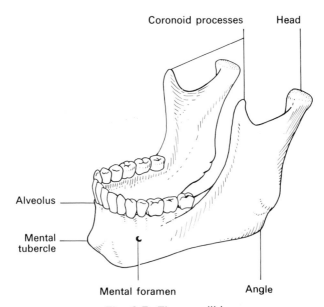

Fig. 3.7. The mandible.

The mandible (Fig. 3.7) is the bone of the lower jaw and carries the lower teeth. It consists of:

(a) a horizontal strong arched body; this carries the teeth and the front of the arch forms the chin,

(b) a vertical ramus, which rises from each end of the arch of the horizontal body and divides at its upper end into (i) a thin pointed coronoid process, to which the temporal muscle is attached; (ii) an articular head which articulates with the joint surface on the inferior surface of the temporal bone to form the temporomandibular joint.

THE TEETH

The first dentition

First dentition
20 teeth
  2 incisors ⎫
  1 canine   ⎬ in each jaw on each side
  2 molars   ⎭

The first dentition teeth are smaller and simpler in form than the permanent teeth. They are already formed in the jaw by the 9th month of intra-uterine life. They erupt (i.e. appear through the gums):

2 lower central incisors at about 6 months,

2 upper incisors at about 8 months,

lateral lower incisors, canine and 1st molar at 15–20 months,

2nd molars at 20–24 months.

Just before the appearance of the corresponding permanent teeth the crowns of the milk teeth break off and the roots are absorbed.

The second dentition

The second dentition teeth begin to form under the milk teeth during intra-uterine life and erupt at periods between 6 years and 25 years. The third molars (wisdom teeth) sometimes do not erupt.

Second dentition
32 teeth
  2 incisors   ⎫
  1 canine   ⎪
  2 premolars ⎬ in each jaw on each side
  3 molars   ⎭

Composition of teeth

Each tooth consists of:

a *crown*—the part that projects into the mouth,

a *neck*—a slightly narrowed ring where the gum is attached,

a *root or roots*—fixed into the alveolus in the bone.

The incisor teeth (Fig. 3.8) have a flat cutting edge to the crown and one root; the canine teeth are long and pointed and have one root; the premolars have two tubercles on the surface of the crown and two roots; the molars have three to four or five tubercles and two to three roots. Opposing teeth are not exactly opposite each other, so that each tooth is in contact with two in the

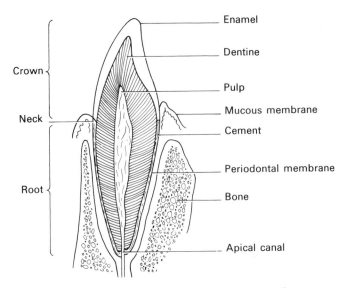

Fig. 3.8. Section through an incisor tooth.

opposite jaw. The upper dental arch is wider than the lower, so that when the jaws are closed the upper incisors and canines overlap the lower incisors and canines.

Each tooth is formed of:

enamel, on the outer surface of the crown,

dentine, beneath the crown and forming the roots,

a *dental cavity*, containing connective tissue, nerves, blood vessels and lymphatics, which enter it through canals in the tip of the root.

The hyoid bone

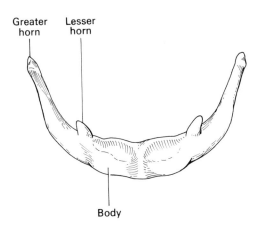

Fig. 3.9. The hyoid bone from the front.

The hyoid bone (Fig. 3.9) is a small bone in the neck. It does not articulate with any other bone. It lies in the midline below the tongue and above the larynx. It consists of a body, two greater horns and two lesser horns.

The following are attached to it:

(a) some of the muscles of the tongue,

(b) part of the pharyngeal wall,

(c) the thyrohyoid membrane, which connects it with the thyroid cartilage of the larynx.

It does not become completely ossified until middle age. It can be broken in strangulation.

4
The Vertebral Column and Thoracic Cage

THE VERTEBRAL COLUMN

The vertebral column is a strong curved mobile pillar which carries the skull, the cage of the thorax, and the upper limbs, transmits the weight of the body to the lower limbs, and protects the spinal cord.

The column (Fig. 4.1) consists of a number of vertebrae, which are connected by intervertebral discs and several ligaments. Each vertebra is composed of spongy bone filled with red marrow and enclosed by a thin layer of compact bone.

```
Bones of the column
   7 cervical vertebrae
   12 thoracic vertebrae
   5 lumbar vertebrae
   sacrum
   coccyx
```

Typical vertebrae and joints
Vertebrae show differences based upon a common pattern. The typical vertebra shows:

a *body*: a thick plate of bone, with slightly curved upper and lower surfaces.

a *vertebral arch*, composed of:

(a) a *pedicle* in front: a bar of bone running backwards from the body, with a notch on the adjacent vertebra forms an intervertebral foramen,

(b) a *lamina* behind: a flat piece of bone running backwards and inwards to fuse with its fellow of the opposite side.

the *vertebral foramen*: a large hole bounded by the body in front, the pedicles at the side, and the laminae at the sides and back.

the *intervertebral foramen*: the hole at the side, between two adjacent vertebrae; through it passes the appropriate spinal nerve.

superior and inferior articular processes: articulating with similar processes on the vertebrae above and below.

transverse process: a laterally projecting bar of bone.

a spine: a process pointing backwards and downwards.

An *intervertebral disc* is a disc attached to the surfaces of the bodies of two adjacent vertebrae: it is composed of an *annulus fibrosus*, a ring of fibrocartilagi-

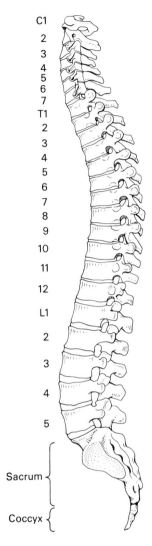

C1
2
3
4
5
6
7
T1
2
3
4
5
6
7
8
9
10
11
12
L1
2
3
4
5
Sacrum
Coccyx

Fig. 4.1. The vertebral column from the side.

nous tissue on the outside, and a *nucleus pulposus*, a semi-fluid substance containing a few fibres and enclosed within the annulus fibrosus.

Ligaments

A number of ligaments connect the vertebrae:
(a) an anterior longitudinal ligament running down the front of the vertebral bodies.
(b) a posterior longitudinal ligament running down the back of the vertebral bodies (i.e. within the vertebral canal),
(c) short ligaments connecting the transverse processes and spines and surrounding the joints on the articular processes.

Cervical vertebrae

Cervical vertebrae (Fig. 4.2) are small, have a thin body, and have transverse processes distinguished by having a foramen (through which passes the vertebral artery) and ending in two tubercles.

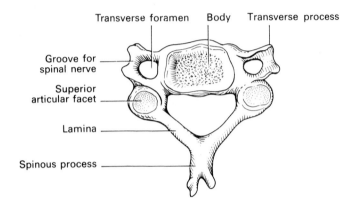

Fig. 4.2. A cervical vertebra from above.

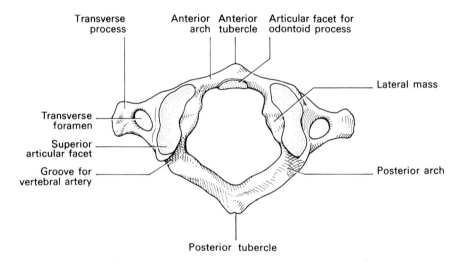

Fig. 4.3. The atlas from above.

The *atlas* (1st cervical vertebra) (Fig. 4.3) differs in having: (a) no body— only a transverse arch of bone in front, (b) an articular surface on the back of this transverse arch for the odontoid process of the axis, (c) articular surfaces above for articulation with the inferior surface of the occipital bone.

The *axis* (2nd cervical vertebra) (Fig. 4.4) differs in having an odontoid process sticking upwards from its body and articulating with the anterior arch

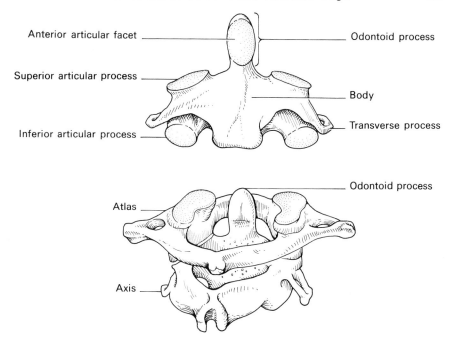

Anterior articular facet

Superior articular process

Inferior articular process

Odontoid process

Body

Transverse process

Odontoid process

Atlas

Axis

Fig. 4.4. The axis (above) from the front, the atlas and axis (below) from the back.

of the atlas. This odontoid process is developmentally the body of the atlas which has become transferred to the axis. It is held in place by short ligaments which connect it to the atlas and the sides of the foramen magnum of the occipital bone.

Clinical feature
Dislocation backwards of the odontoid process can cause severe injuries to the cervical part of the spinal cord, which is immediately behind it. Death or complete paralysis below the lesion may occur.

Thoracic vertebrae

These are typical vertebrae (Fig. 4.5). In addition to the usual features they show:
(a) an articular surface on the side of the body for articulation with the head of a rib,
(b) an articular surface on the transverse process for articulation with the tubercle on a rib.

They become larger from above downward as they have to carry an increasingly greater weight, and the 12th is a massive vertebra resembling a lumbar vertebra.

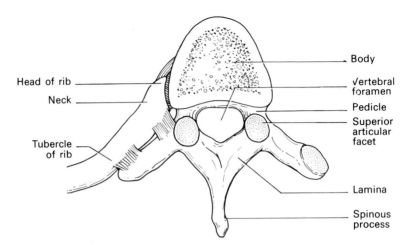

Fig. 4.5 A thoracic vertebra from above, showing articulation with a rib.

Lumbar vertebrae

The lumbar vertebrae (Fig. 4.6) are massive bones with thick strong lateral processes and spines.

The *vertebral canal* is formed by the successive vertebral foramina and by the intervertebral discs and ligaments connecting them. It contains:
(a) the spinal cord, as far down as the 1st or 2nd lumbar vertebra,
(b) the spinal nerves as they leave or enter the spinal cord,
(c) blood vessels,
(d) the meninges (coverings of the cord).

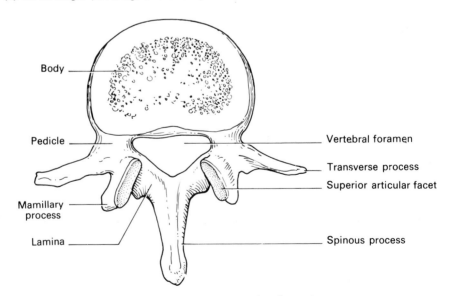

Fig. 4.6. A lumbar vertebra from above.

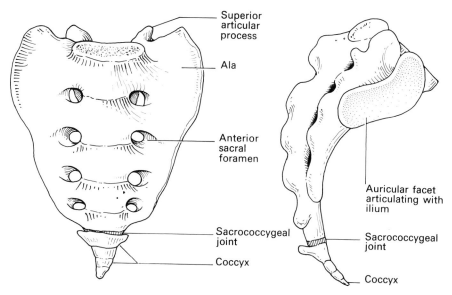

Superior articular process

Ala

Anterior sacral foramen

Auricular facet articulating with ilium

Sacrococcygeal joint

Coccyx

Sacrococcygeal joint

Coccyx

Figs 4.7 & 4.8. The sacrum from the front and side.

The sacrum (Fig. 4.7) is composed of five vertebrae fused together. It is a curved wedge of bone, showing:

(a) a smooth concave surface anteriorly, which forms the back of the pelvic cavity,

(b) a rough convex posterior surface to which are attached ligaments and parts of the erector spinae muscle and gluteus maximus muscle,

(c) an articular surface at each side for articulation with the ilium,

(d) a small articular surface below for articulation with the coccyx,

(e) four anterior and four posterior sacral foramina through which pass the anterior and posterior branches of the sacral nerves.

The coccyx

The coccyx (Fig. 4.8) is a small triangular bone, formed of four coccygeal bones fused together. It articulates above with the sacrum and forms part of the posterior wall of the pelvis.

Clinical feature
Coccydynia is a painful condition of the coccyx which sometimes follows a fall on the buttocks.

Length and shape of the vertebral column

The length of the vertebral column is approximately the same for all people of about average height: 70 cm for men, 60 cm for women. The intervertebral discs form about one-fifth of the total height.

The vertebral column seen from in front is usually vertical. There may be a slight deviation to one side or the other in the thoracic region, with compensatory curves in the opposite direction in the cervical and lumbar regions.

Before birth the column shows a concavity forwards. With the development of the head and the adoption of a vertical posture, convexities forwards appear in the cervical and lumbar regions. In its final form therefore the spinal column shows:

    cervical region: curve forwards,
    thoracic region: curve backwards,
    lumbar region: curve forwards,
    sacral and coccygeal region: curve backwards above and forwards below.

Movements of the vertebral column

The vertebral column can be flexed, extended, rotated and moved laterally. These movements are made possible by:

(a) small movements between adjacent vertebrae,

(b) alterations in the intervertebral discs, which are compressible and expansible.

The shape of the atlas and axis make possible the nodding and rotatory movements of the head.

The movements are produced by the action of a large number of muscles attached to the vertebral column along its length. The strong erector spinae group of muscles, extending from the sacrum to the head, fills in the gaps on either side of the spines of the vertebrae and has a powerful extensor action. The other actions are carried out by many muscles, including the sternomastoid muscles, deep muscles in the neck, muscles attached to the ribs, and the muscles of the abdominal wall.

Clinical features

Scoliosis is a lateral curvature of the vertebral column. *Lordosis* is an extension deformity so that the patient stands with his trunk pushed forwards. *Kyphosis* is a flexion deformity so that the patient stands with his head bent downwards. *Collapse of the body of a vertebra* can be due to a fracture of the body or to disease (usually cancer or tuberculosis); the vertebral body becomes wedge-shaped with the thin edge forwards; untreated, it causes hunchback, which is an extreme form of kyphosis.

Backache is commonly due to minor asymmetry of the vertebral column, especially in the lumbosacral region. A *prolapsed intervertebral disc* is the result of a rupture of the annulus fibrosus of an intervertebral disc. The nucleus pulposus of the disc is extruded through the opening and can press on a spinal nerve or less commonly on the spinal cord. When this happens in the cervical region the patient has an acutely stiff neck and pain radiating down the arms.

When it happens in the lumbar region the patient gets a low back pain (lumbago) or pain down a leg (sciatica).

THE THORACIC CAGE

Thoracic cage
  sternum
  ribs and costal cartilages
  thoracic part of vertebral column

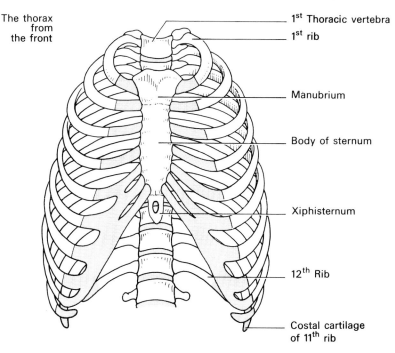

The thorax from the front

1st Thoracic vertebra
1st rib
Manubrium
Body of sternum
Xiphisternum
12th Rib
Costal cartilage of 11th rib

Fig. 4.9. The thoracic cage formed of sternum, ribs and thoracic part of vertebral column.

Sternum

The sternum is a flat, blade-shaped, subcutaneous bone situated in the midline in the front of the chest. It is composed of three parts:
  manubrium,
  body,
  xiphoid process
  The *manubrium*, the upper part, is about 5 cm long and runs downwards and slightly forwards to meet the body at an easily felt angle called the sternal angle.

The suprasternal notch is a midline notch in its upper border. The clavicles articulate with it on either side at its laterosuperior angle. The costal cartilage of the 1st rib articulates with the lateral border and that of the 2nd with the sternal angle.

The body is a long, flat, slightly ridged bone, with articulations along its lateral borders for the 3rd to 7th costal cartilages.

The *xiphoid process* is a small plate, sometimes perforated by a hole or divided into two. The cartilage with which it starts ossifies much later than the rest of the bone; the joint between it and the body does not ossify until middle age.

The *posterior surface* of the sternum is smooth and slightly concave. It is in contact with the structures of the superior and anterior mediastinum (see p. 126): the thymus, the heart within the pericardium, the lungs within the pleurae.

The ribs

The ribs are twelve pairs of bones which pass round the wall of the chest, articulating behind with the vertebral column and in front, through the costal cartilages, with the sternum (Fig. 4.10).

Each rib has:

a *head*, which articulates with the vertebral column,

a *neck*, short and passing backwards and slightly laterally,

a *tubercle*, a short projection articulating with the lateral process of a vertebra,

a *shaft*, the long part of the bone, running forwards and downwards round the chest wall to end in front in a costal cartilage.

The ribs have external and internal surfaces. The upper border is rounded. The lower border has a groove in which run the intercostal vessels and nerves. The space between two adjacent ribs is occupied by the external and internal intercostal muscles, which are attached to the two bones.

There are individual differences between the ribs, especially:

1st rib: short, with its surfaces facing mostly upwards and downwards.

11th and 12th ribs: rudimentary, short, pointed.

The *costal cartilages* connects ribs 1–10 with the sternum. The costal cartilages of ribs 7–10 are connected by a common piece of cartilage with the junction of the sternal body and xiphoid cartilage. The costal cartilages of ribs 11–12 are short, thick and pointed, do not reach the sternum, and end in the muscles of the abdominal wall.

The *thoracic inlet*, at the upper end of the thorax, is kidney-shaped, about 10 cm wide and about 5 cm from front to back. Its boundaries are:

1st thoracic vertebra,

1st rib and its costal cartilage,

upper border of manubrium.

It faces downwards and forwards. It contains the apices of the lungs, and through it pass the trachea, oesophagus, common carotid arteries, and internal jugular vein.

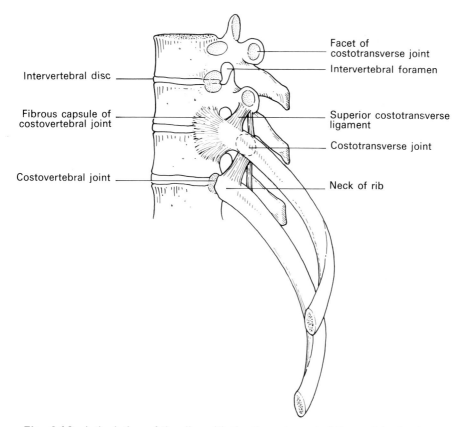

Intervertebral disc

Fibrous capsule of costovertebral joint

Costovertebral joint

Facet of costotransverse joint

Intervertebral foramen

Superior costotransverse ligament

Costotransverse joint

Neck of rib

Fig. 4.10. Articulation of the ribs with the thoracic part of the vertebral column.

The *thoracic outlet*, at the lower end, is large, irregular and bounded by:
    12th thoracic vertebra,
    lower ribs and costal cartilages,
    the xiphisternal joint,
It is closed below by the diaphragm.

Movements of the thoracic cage

The size of the thoracic cavity is enlarged in inspiration and reduced in expiration.

In *inspiration*: (a) the muscle of the diaphragm contracts and pulls down its central tendon, thus increasing the length of the cavity; (b) contraction of the external intercostal muscles rotates the ribs outwards and upwards, thus increasing the width of the cavity.

In *expiration* there is an elastic recoil of the cage to its resting position and an elevation of the central tendon of the diaphragm.

In quiet respiration the 1st rib does not move and the 2nd very little, and the 11th and 12th ribs are fixed by the abdominal muscles.

The size of the thoracic cavity is increased by extension of the vertebral column and decreased by flexion of it.

Clinical features

Fractures of ribs may be due to a direct blow or to a severe compression of the chest. The fracture of a single rib does not usually produce complications, but when several ribs are broken the chest wall may be driven in and the heart or lungs damaged. A *fracture of the sternum* may be the result of the same injuries that cause fracture of the ribs or be due to a forcible flexion injury of the thoracic spine.

5
The Bones of the Limbs

BONES OF THE ARM

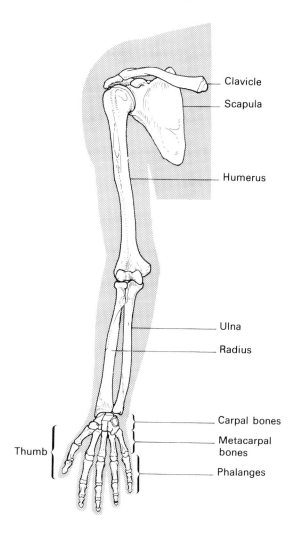

Clavicle

Scapula

Humerus

Ulna

Radius

Carpal bones

Metacarpal bones

Phalanges

Thumb

Fig. 5.1. The bones of the arm and hand from the front.

Bones of arm
  scapula
  clavicle
  humerus
  radius
  ulna
  carpal bones
  metacarpals
  phalanges

Scapula (shoulder-blade)

The scapula (Figs 5.2, 5.3) is a flat triangular bone forming part of the shoulder girdle. It has two surfaces—anterior and posterior, and three borders—superior, lateral and medial. Its anterior surface is slightly concave and lies against the posterior wall of the thorax. The posterior surface is divided into two areas by the spine of the scapula, a ridge of bone, palpable through the skin, which runs across the width of the scapula to end laterally in the acromion process, a thick bar of bone lying immediately above the shoulder joint. The acromion articulates with the lateral end of the clavicle. A small pointed corocoid process sticks forwards from the upper border of the scapula, projecting just below the clavicle. The glenoid cavity, at the upper end of the outer border of the scapula, articulates with the head of the humerus to form the shoulder joint.

The scapula is attached to the head, trunk and arm by a number of muscles. In movements of the shoulder joint it glides over the posterior surface of the chest wall.

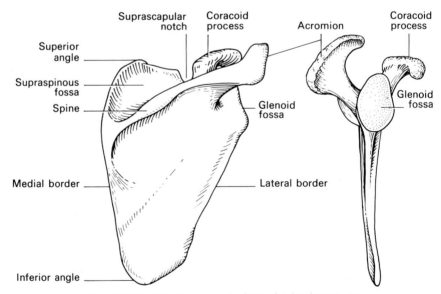

Figs 5.2 & 5.3. The scapula from the back and side.

Clinical feature
Fracture of the scapula is uncommon because it is largely protected by covering muscles, and lies flat against the chest wall.

Clavicle (collar-bone)

The clavicle is a long, slightly S-shaped bone attached at its medial end to the manubrium of the sternum and at its lateral end to the acromion process of the scapula. Its presence and position enables the arm to hang away from the body, and gives a big range of movements to the shoulder joint.

Clinical features
Fracture of the clavicle can occur at all ages, but is most common in children and young adults. It is usually the result of a fall on the shoulder. The fracture usually occurs about the middle of the bone.

Humerus

The humerus is a long bone with
    a head (the upper end),
    a shaft,
    a lower end.

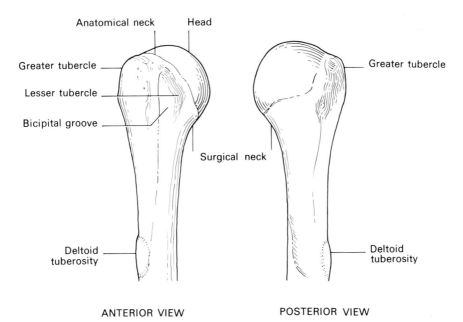

ANTERIOR VIEW   POSTERIOR VIEW

Fig. 5.4. The upper end of the humerus.

The *head* (Fig. 5.4) is nearly hemispherical and articulates with the glenoid cavity of the scapula. The *anatomical neck* is the narrow groove immediately below the head. The *greater tuberosity* (in front) and the *lesser tuberosity* (behind) are two elevated prominences for the attachment of muscles. The *surgical neck* is the upper end of the shaft immediately below the tuberosities.

The *shaft* is a cylindrical bar of bone. The deltoid tuberosity is a V-shaped ridge on its lateral aspect half way down, for the insertion of the deltoid muscle. The spiral groove is a groove at the back of the shaft down which the radial nerve runs. Muscles are attached to the rest of the shaft.

The *lower end* (Fig. 5.5) is wide, flattened anteroposteriorly, has a lateral epicondyle on the lateral aspect and a medial epicondyle at the medial aspect, both for the attachment of muscles. The capitulum is a rounded eminence with a joint surface which articulates with the head of the radius; the trochlea is the articular surface, medial to the capitulum, for articulation with the upper end of the ulna.

ANTERIOR VIEW   POSTERIOR VIEW

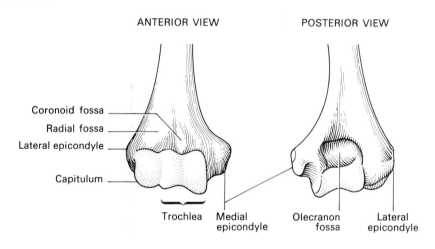

Coronoid fossa
Radial fossa
Lateral epicondyle
Capitulum

Trochlea   Medial   Olecranon   Lateral
           epicondyle   fossa   epicondyle

LATERAL VIEW

Lateral   **Fig. 5.5.** The lower end
epicondyle   of the humerus.

Clinical features

Fractures of the humerus are most common at:

(a) the surgical neck,

(b) the shaft; in a fracture of the shaft the radial nerve, running in the radial groove, may be injured,

(c) the lateral or medial epicondyle.

Radius and ulna

The *radius* is the bone on the outer side of the forearm (Fig. 5.6). It has:

    an *upper end* with

(a) the head, which articulates with the capitulum of the humerus,

(b) a neck,

(c) a tuberosity, to which is attached the tendon of the biceps muscle.

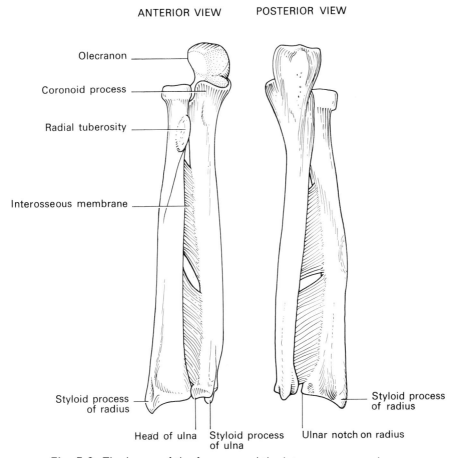

Fig. 5.6. The bones of the forearm and the interosseous membrane.

a *shaft*, to which various flexor and extensor muscles of the forearm are attached.

a *lower end*, with a pointed styloid process and an articular surface for some of the wrist bones and an articular surface for the lower end of the ulna.

The *ulna* is the long bone on the inner side of the forearm. It shows:

an *upper end* with

(a) the olecranon, with a pointed process which lies behind the lower end of the humerus,

(b) the coronoid process, a pointed process in front,

(c) articular surfaces on these processes for the lower end of the humerus and on the outer side for the head of the radius.

a *shaft*, which tapers and to which flexor and extensor muscles of the forearm and hand are attached.

the *lower end* with

(a) a small styloid process,

(b) an articular surface for the lower end of the radius,

(c) an articular surface which is separated from the bones of the wrist by a pad of cartilage.

The *interosseus membrane* is a sheet of fibrous tissue, which is attached to adjacent edges of the radius and ulna and fills the gap between them. It provides attachment for muscles both in front and behind.

Clinical features

Fractures can occur of (a) the olecranon, (b) the neck of the radius, (c) the shafts of both bones, (d) the lower ends of both bones.

Colles' fracture is a fracture of the lower end of the radius (the lower end being pushed backwards, giving to the wrist the appearance of a dinner-fork) combined usually with a fracture of the styloid process of the ulna.

Carpus (wrist)

The carpus (Fig. 5.7) consists of eight small irregular bones arranged in two rows:

proximal row (lateromedially): scaphoid, lunate, triquetrum, pisiform.
distal row (lateromedially): trapezium, trapezoid, capitate, hamate.

The carpus articulates above with the radius and ulna and below with the metacarpals.

Clinical feature

Fracture of the scaphoid is usually due to a fall on the hand or to a violent twist of the wrist. The fracture does not heal quickly if the fracture cuts off the blood supply to one half, and a long period of immobilization in plaster is necessary.

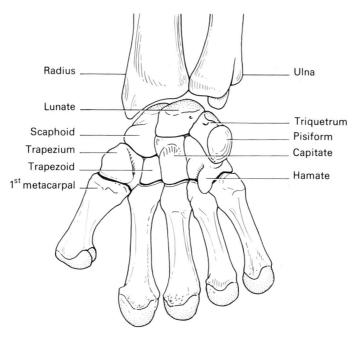

Fig. 5.7. The bones of the wrist from the front.

Metacarpals

The metacarpals (Fig. 5.8) are five bones in the hand. They have:

a base, which articulates with the carpus,

a shaft,

a head, a rounded end which articulates with the first of the phalanges of the appropriate digit.

The metacarpal of the thumb is particularly short and strong.

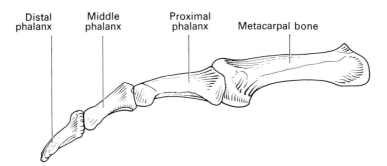

Fig. 5.8. The metacarpal bone and phalanges of one finger.

Phalanges

The thumb has two phalanges, the fingers three. They diminish in size towards the end of the digit. On the distal phalanx is a roughened area for the pad of the finger.

BONES OF THE LEG

> *Bones of leg*
>   hip (innominate) bone
>   femur
>   patella
>   tibia
>   fibula
>   tarsal bones
>   metatarsals
>   phalanges

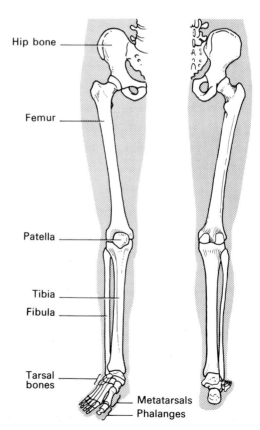

Hip bone

Femur

Patella

Tibia

Fibula

Tarsal bones

Metatarsals

Phalanges

Fig. 5.9. The bones of the leg and foot from the front and back.

The pelvis

The pelvis is formed by:
    the hip bone in front and at the sides,
    the sacrum and coccyx behind.
It is a bony arch through which the weight of the body is transferred to the legs.

The *false pelvis* is the large upper part of the pelvis between the two iliac bones of the hip bone. In life the space is largely occupied by bowel.

The *true pelvis* is the lower, smaller part of the pelvis. The pelvic brim marks the division of false from true pelvis. The true pelvis contains:

in men	*in women*
bladder,	bladder,
prostate gland,	uterus, uterine tubes, ovaries, vagina,
rectum,	rectum,

    the blood vessels, lymph vessels, nerves to supply these organs,
    connective tissue.

Clinical features

Fractures of the pelvis may be complicated by injuries to any of the organs it contains. The size of the true pelvis is important in childbirth; it has to be big enough to let the head (which is the largest single part of the fetus) pass through it.

The hip (innominate) bone

The hip bone (Figs 5.10, 5.11) is a large, thick, strong, irregularly shaped bone. It articulates behind with the sacrum and in front with its fellow bone of the opposite side.

It consists of three bones which have fused into one.

Ilium. The ilium forms the upper and posterior parts of the hip bone. The iliac crest is its upper border; the crest ends in front in the anterior superior iliac spine (which can be felt in the living person) and behind in the posterior superior iliac spine. Its inner surface is smooth and hollowed out, and provides attachment for the iliac muscle. Its outer surface is ridged and provides attachments for the gluteal muscles.

Ischium. The ischium is below the ilium, has an ischial tuberosity, a large thick mass of bone in the gluteal region, and articulates in front with the pubic bone.

Pubic bone.   The pubic bone forms the front of the hip bone and articulates with its fellow to form the pubic arch. It shows:

(a) a joint surface for the other pubic bone; the joint is called the symphysis pubis,

(b) an upper part marked by a prominent pubic tubercle,

(c) a lower part which runs backwards and downwards to join the ischium.

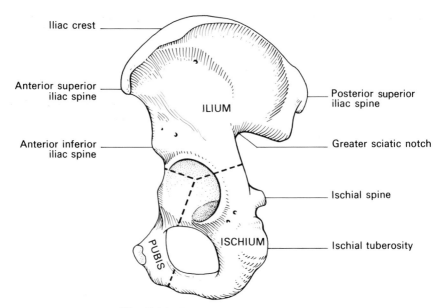

Fig. 5.10. The hip bone, lateral aspect.

Iliac crest

Anterior superior iliac spine

Anterior inferior iliac spine

ILIUM

Posterior superior iliac spine

Greater sciatic notch

Ischial spine

Ischial tuberosity

ISCHIUM

PUBIS

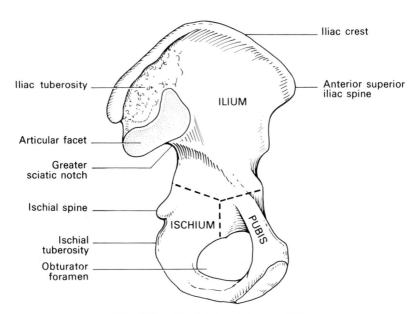

Fig. 5.11. The hip bone from within.

Iliac tuberosity

Articular facet

Greater sciatic notch

Ischial spine

Ischial tuberosity

Obturator foramen

Iliac crest

Anterior superior iliac spine

ILIUM

ISCHIUM

PUBIS

The *obturator foramen* is a hole bounded in front by the pubic bone, and behind by the ischium. In life it is filled in by an interosseous membrane, to which muscles are attached.

The hip bone has three articulations:

(a) with the sacrum by a large articular surface on each ilium,

(b) with the femur by the acetabulum, a hollowed-out depression in which all the three bones have a share. The head of the femur fits into it,

(c) the hip bone of the opposite side at the symphysis pubis.

The femur

The femur consists of:

    an upper end,
    a shaft,
    a lower end.

The *upper end* (Fig. 5.12) consists of:

(a) a head: a rounded mass directed inwards and upwards; it is smooth and covered with cartilage except at the fovea, a little pit to which is attached a short ligament which connects the head to a roughened area on the acetabulum of the hip bone,

(b) a neck: a shaft of bone directed downwards and laterally, connecting the head and the shaft,

(c) a greater trochanter laterally and a lesser trochanter medially: eminences for the attachment of muscles.

The *shaft* is a long bar of bone, tapering slightly towards the middle. Most of its surface is smooth and has muscles attached to it. Posteriorly the *linea aspera*

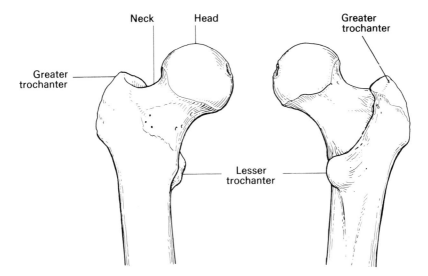

Fig. 5.12. The upper end of the femur from the front and back.

is a double ridge of bone, running downwards from the trochanters above and widening out below to enclose a smooth area.

The *lower end* consists of large medial and lateral condyles and an area of bone between them. The condyles have articular surfaces for the tibia below and the patella in front.

Clinical feature
A *fracture of the neck of the femur* is the commonest of the fractures of the femur. It is particularly liable to occur in old people as the result of a fall. The fracture does not heal quickly as it is liable to deprive the head of the femur of much of its blood supply. To aid healing and to enable the patient to become mobile as soon as possible, this fracture is usually treated by driving a steel pin through the greater trochanter into the head. The patient is thus able to get out of bed and start walking.

The patella

The patella (Fig. 5.13) is roughly triangular, with rounded angles and an apex pointing downwards. It is the largest of the *sesamoid bones*, which are bones formed in the tendons of muscles for mechanical purposes. The patella is formed in the tendon of the quadriceps muscles of the thigh. It glides over the articular surface on the front of the lower end of the femur, acting as a kind of movable lever and giving the quadriceps muscles some improved pulling action.

Clinical features
The patella can be fractured by a direct blow. It does not heal readily and even when healed, there can be some impairment of its articulation with the femur.

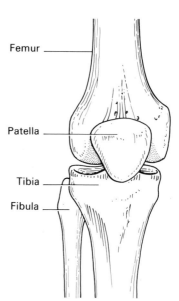

Fig. 5.13. The bones of the knee joint from the front.

Surgical removal is sometimes performed; loss of the patella does not lead to much mechanical disadvantage.

Tibia and fibula

The tibia and fibula are the bones of the leg below the knee.

TIBIA

The tibia is on the medial side and transmits the weight of the body. It consists of:

an upper end,
a shaft,
a lower end.

The *upper end* is widened transversely and has a superior articular surface on each of its condyles, medial and lateral. There is a non-articular rough area between the articular surfaces for the attachment of ligaments. The semilunar cartilages rest on the upper surface of the tibia, separating it from the condyles of the femur. The upper end of the fibula is attached to an articular surface on the lateral condyle.

The *shaft* is triangular in section and its anterior border forms the prominent palpable shin. It narrows to about its midpoint and then enlarges.

The *lower end* shows (a) the medial malleolus, a pointed eminence, on the inner aspect of the ankle, (b) an articular surface for the lower end of the fibula, and (c) articular surface below and medially for the talus.

FIBULA

The fibula is a long slim bone on the lateral aspect of the leg. It shows:

an upper end, which articulates with the lateral condyle of the tibia
a shaft
a lower end, which shows: (a) the lateral malleolus of the ankle, (b) articular surface for the lower end of the tibia, and (c) articular surface for the talus.

The tibia and fibula are joined together above and below by immovable joints. An *interosseous membrane* is attached to the shafts of the bones and occupies the space between them; it serves for the attachment of muscles.

Clinical feature

Pott's fracture is a fracture-dislocation at the ankle; either (a) the lateral malleolus is broken, or (b) a fracture of the lateral malleolus is associated with either a tear of the medial ligament of the ankle joint or less commonly with a fracture of the medial malleolus of the tibia, as the result of external rotation and lateral displacement of the talus at the time of the accident.

The foot

> *Bones of foot*
>   talus
>   calcaneum
>   navicular bone
>   cuboid bone
>   3 cuneiform bones
>   metatarsals
>   phalanges

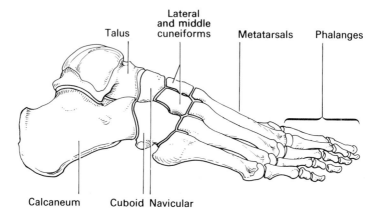

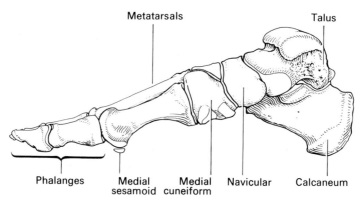

Fig. 5.14. The bones of the foot lateral aspect (above) and medial aspect (below).

TARSUS

The *talus* is an irregularly shaped bone. It receives the weight of the body transmitted through the tibia. It articulates:

　　above, with the tibia
　　medially, with the medial malleolus

laterally, with the lateral malleolus
below, with the calcaneum
in front, with the navicular bone.

The *calcaneum* is a strong, thick, irregularly shaped bone, which posteriorly forms the heel. The tendon of Achilles is attached to it posteriorly. It articulates:
above, with the talus
below, with the cuboid bone

The *navicular bone* is a flat bone, which articulates:
behind, with the talus
in front, with the three cuneiform bones.

The *cuboid bone* is approximately cuboid and lies on the outer side of the foot. It articulates:
behind, with the talus
medially, with the navicular bone and the lateral cuneiform bone
in front, with the 4th and 5th metatarsals.

The three *cuneiform bones* lie in a row between the cuboid bone on the lateral aspect and the side of the foot medially. They articulate:
behind, with the talus
in front, with 1st, 2nd and 3rd metatarsals
the lateral cuneiform bone laterally, with the cuboid bone.

METATARSALS

There are five metatarsals, one for each toe. Each has a base, a shaft and a head. The 1st metatarsal is short, thick and strong. The 1st, 2nd and 3rd articulate with the cuneiform bones, the 4th and 5th with the cuboid bone. Each metatarsal articulates with the appropriate proximal phalanx.

PHALANGES

The big toe has two phalanges, the others have three. Each phalanx has a shaft and two ends; but the middle phalanges are short and the distal phalanges small.

6

Joints and Movements

> *Joints*
>   fibrous
>   cartilaginous
>   synovial

FIBROUS JOINTS

(a) *suture*: a thin strip of fibrous tissue separating two interlocking bones, it is found only in the skull, and ossifies in later life.
(b) in some joints, e.g. the lower tibiofibular joint, the bones are held together by a fibrous ligament.

CARTILAGINOUS JOINTS

Bones are separated by cartilage, e.g. at the junction of epiphysis and diaphysis (shaft) in a developing bone, between the bodies of vertebrae, at the manubrio-sternal joint, at the symphysis pubis.

SYNOVIAL JOINTS

Most joints are synovial (Fig. 6.1). They are composed of:
(a) *cartilage*: the bony surface within the joint are covered with smooth, moist, silvery-blue hyaline cartilage; it has no nerves and no blood supply,
(b) *capsule*: formed of fibrous tissue, encloses the joint as within a bag, and is attached to the periosteum around the periphery of the joint area,
(c) *synovial membrane*: lines the inner surface of the capsule,
(d) *synovial fluid*: a small amount of fluid secreted by the synovial membrane to provide a lubricant for the joint,
(e) *ligaments*: fibrous thickenings reinforcing the capsule on the outside and attached to adjacent bone.

Some joints (e.g. the knee joint) have pads of fibrous tissue within them.

Joints are supported and strengthened on the outside by muscles and tendons. The stability of a joint is partly dependent upon the shape of the bones

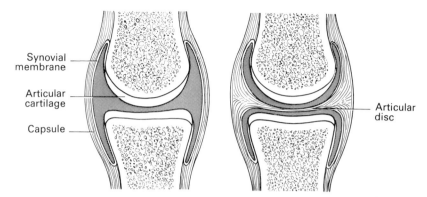

Synovial membrane

Articular cartilage

Capsule

Articular disc

Fig. 6.1. The basic structures of a joint containing an articular disc.

involved, but mainly on the tone and strength of the muscles surrounding it and acting upon it and on their tendons. The many blood vessels around a joint disperse the considerable heat that repeated movement produces.

Types of joints

   plane *joints*: flat or nearly flat surfaces, e.g. the intertarsal joints, the acromioclavicular joint
   hinge joints: at which movements of flexion and extension take place, e.g. the elbow
   saddle joints: the articular surface of one bone has the shape of a saddle, e.g. 1st metacarpal-trapezium joint in the wrist
   ball and socket joint: the ball of one bone fits into the cup of the other, e.g. the hip joint
   condyloid joint: a hinge joint at which some lateral movement can take place, e.g. the temporomandibular joint
   pivot joint: one bone rotates within a ring made of a ligament or partly of bone, partly of ligament, e.g. the radio-ulnar joints, the joint between the odontoid process and the atlas

---

Movements of joints

Gliding: the movement of one surface over another as at a plane joint.
Flexion: decreasing the angle of a joint, e.g. bending the elbow.
Extension: increasing the angle of a joint, e.g. straightening the elbow.

Dorsiflexion of foot or toes: bending the foot or toes upwards.
Plantar flexion of foot or toes: bending the foot or toes downwards.

Abduction: movement of the part away from the midline of the body, e.g. lifting the arm away from the side.

Adduction: movement of the part towards the midline of the body, e.g. bringing the arm to the side.

Rotation: movement of the part round its own longitudinal axis, e.g. moving the palm forwards and backwards with the arm fully extended is achieved by rotation at the shoulder joint.

Circumduction: a combination of flexion, abduction, extension and adduction in one movement, e.g. swinging the extended arm round in a circle.

Clinical features

Arthritis is inflammation or degeneration of a joint. *Osteo-arthritis* is an arthritis of a big joint (such as the hip) as a result of weight-bearing or repeated jarring; degeneration of cartilage and bone occurs. *Rheumatoid arthritis* is an autoimmune disease producing inflammatory and degenerative changes in many small joints and is often severely crippling.

Dislocation is the forcing of the bones of a joint out of their normal position. A *fracture-dislocation* is a combination of a fracture with a dislocation.

IMPORTANT JOINTS

Temporomandibular joint (Fig. 6.2)

Type: condyloid
Bones: articular fossa of temporal bone,
        head of mandible.
Special feature: contains an articular disc of fibrous tissue between the bones.
Main muscles: muscles of mastication (masseter, temporalis, pterygoids).

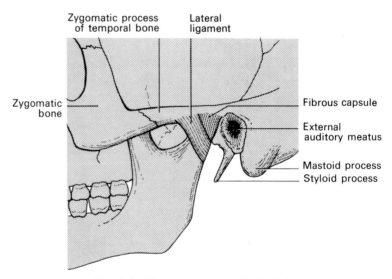

Fig. 6.2. The temporomandibular joint.

Movements: opening and closing the mouth; rotary (grinding) movements; protrusion of lower jaw.

Joints of vertebral column

Bones: articular surfaces on the articular processes of adjacent vertebrae form individual joints.

Special feature: there is a small joint between the back of the anterior arch of the atlas and the front of the odontoid process of the axis.

Muscles : extension—erector spinae,

flexion—long muscles of the neck, rectus abdominis, muscles of anterolateral abdominal wall, psoas.

Movements: are mainly flexion and extension. The range of movement at any one joint is small, but the accumulation of them all produces bending and stretching through a wide range. Flexion occurs mainly in the cervical region and least in the thoracic region. Extension occurs mainly in the lumbar region. The intervertebral discs permit one vertebra to rock upon another, one edge of the disc being compressed as the opposite expands.

Other movements are rotation of the vertebral column around its long axis and bending from side to side.

The particular shape of the atlas and axis and their joints enable nodding and rotatory movements to occur.

Joints of the arm

ACROMIOCLAVICULAR JOINT (Fig. 6.3)

Type: plane.
Bones: acromion of the scapula,

lateral end of a clavicle.

Special features: (a) a small incomplete disc of fibrocartilage is usually present between the bones, (b) a coracoclavicular ligament strengthens the joint by attacking the clavicle to the coracoid process of the scapula immediately below it.

Movements: short gliding movements.

STERNOCLAVICULAR JOINT

Type: modified ball and socket.
Bones: clavicular notch on manubrium of the sternum,

medial end of clavicle and adjacent part of 1st costal cartilage.

Special feature: a disc of fibrocartilage is present between the bones.

Main muscles: sternomastoid,

pectoralis major.

Movements: slight ball and socket movements.

SHOULDER JOINT

Type: ball and socket.
Bones: glenoid cavity of the scapula,
        head of the humerus.
Special features: (a) the glenoid cavity is deepened by a fibrous ring attached to its edge; (b) the capsule is loose and enables a wide range of movements to occur; (c) the tendon of the long head of biceps passes through the joint and over the head of the humerus, helping to maintain the stability of the joint; (d) the strength of the joint (obvious when a mother yanks her child out of danger) is mainly in the muscles in front, behind, above and below it.
Main muscles: pectoralis major,   deltoid,
                 latissimus dorsi,   long head of triceps,
                 supraspinatus,   infraspinatus.
Movements: forwards (flexion) and backwards (extension),
            abduction and adduction,
            rotation.
Movements at the shoulder are combined with movements of the scapula over the surface of the thorax.

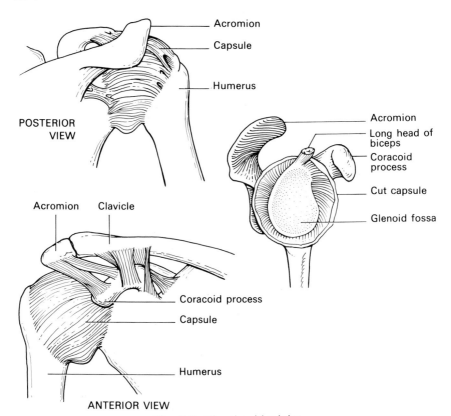

Acromion
Capsule
Humerus

POSTERIOR
VIEW

Acromion
Long head of biceps
Coracoid process
Cut capsule
Glenoid fossa

Acromion   Clavicle

Coracoid process
Capsule

Humerus

ANTERIOR VIEW
Fig. 6.3. The shoulder joint.

Clinical feature
Dislocation of the shoulder, a common injury, is usually the result of falling on the hand with the arm abducted. The head of the humerus is forced through a relatively weak part of the capsule inferiorly and comes to lie, usually in front of the glenoid cavity, rarely behind it.

ELBOW JOINT (Fig. 6.4)

Type: hinge.
It is continuous with the joint at the upper ends of the radius and ulna.

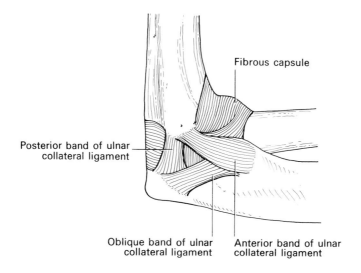

Fibrous capsule

Posterior band of ulnar collateral ligament

Oblique band of ulnar collateral ligament

Anterior band of ulnar collateral ligament

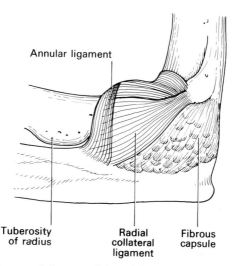

Annular ligament

Tuberosity of radius

Radial collateral ligament

Fibrous capsule

Fig. 6.4. The elbow joint; medial aspect (above) lateral aspect (below).

Bones: trochlea and capitulum at the lower end of the humerus,
          upper end of ulna,
          upper end of radius.
Special feature: The joint is supported and strengthened by the tendon of biceps
in front and the tendon of triceps behind.
Main muscles: flexors—biceps, brachialis,
                extensors—triceps.
Movements: flexion—bending the elbow,
             extension—straightening the elbow.

UPPER AND LOWER RADIO-ULNAR JOINTS

(a) *Upper radio-ulnar joint*. *Type*: pivot. The head of the radius rotates on the
lower end of the humerus and upper end of the ulna in a joint which is continuous with the elbow joint. The neck of the radius is encircled by a strong
ligament attached to the anterior and posterior margins of the radial notch on
the ulna.
(b) *Lower radio-ulnar joint*. *Type*: pivot. The lower end of the radius rotates on
the lower end of the ulna.

An *interosseous membrane* attaches the shaft of the radius to the shaft of the
ulna and provides surfaces in front and behind for the attachment of muscles of
the forearm.
Movements: supination—turning the palm forwards,
             pronation—turning the palm backwards.

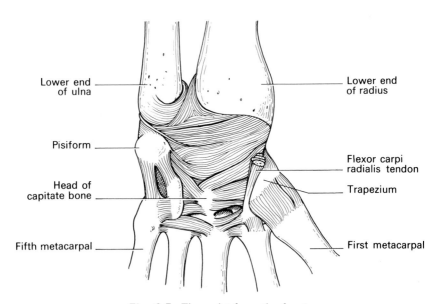

Fig. 6.5. The wrist from the front.

WRIST JOINT

Type: condyloid.
Bones: lower end of radius,
  lower end of ulna, separated from the joint by a piece of cartilage,
  proximal row of carpal bones (Fig. 6.5).
Main muscles: flexors and extensors of hand and fingers.
Movements: flexion and extension,
  abduction and adduction.

CARPAL AND PHALANGEAL JOINTS

Joints between:
  the distal row of carpal bones and the metacarpals,
  metacarpals and proximal phalanges,
  phalanges.
Main muscles: flexors and extensors of the fingers.
Movements: flexion and extension.

Joints of the pelvis

SYMPHYSIS PUBIS

Bones: the symphyseal surfaces of the two pubic bones in the midline anteriorly.
Special features: a fibrocartilaginous disc separates the bones; the ends of the bones are covered by hyaline cartilage; there is no synovial cavity (Fig. 6.6).
Movements: none.

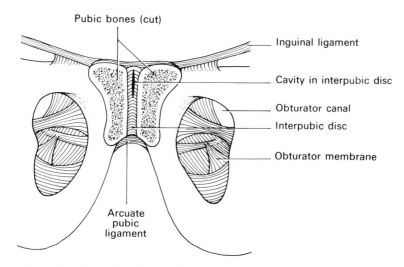

Fig. 6.6. The pubic joint with a section through the pubic bones.

Clinical features

In pregnancy the joint becomes softened and the ligaments surrounding it relax. During childbirth the joint widens out a little to assist the passage of the fetal head through the pelvis.

SACRO-ILIAC JOINT

Bones: the sacrum articulates on each side with an iliac bone at their auricular surfaces.

Special features: the weight of the body is transmitted from the sacrum through the joint to the pelvic bone and lower limbs. Strong anterior, posterior and interosseous ligaments bind the bones together. The bony surfaces are separated by thick cartilage, which can ossify in old age.

Movement: slight rocking movement is possible. Greater movement is possible in women during pregnancy.

Joints of the leg

HIP JOINT

Type: ball and socket.
Bones: head of femur,
        acetabular fossa of innominate bone.
Special features: the capsule is strengthened by ligaments passing from the pubis, ischium and ilium to the femur, the iliofemoral ligament in front being particularly strong; the acetabulum is deepened by a fibrocartilaginous rim attached to its edge (Figs 6.7, 6.8).

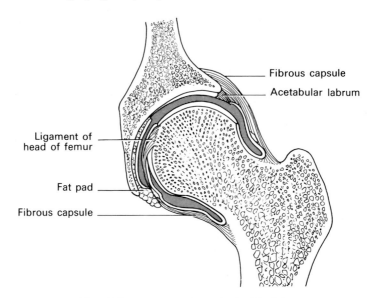

Fig. 6.7. A section through the hip joint.

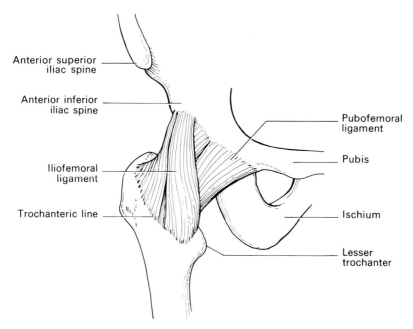

Fig. 6.8. The ligaments of the hip joint from the front.

Main muscles: psoas,   iliacus,
                      rectus femoris,   sartorius,
                      adductors,   hamstrings.
                      gluteals,
Movements: flexion and extension,
                      abduction and adduction,
                      rotation.

Clinical features

In *congenital dislocation of the hip* the head of the femur is displaced out of the acetabulum. The cause of this abnormality is not known. It is rarely complete at birth, but becomes complete as the baby becomes active. If untreated, the child walks with a waddle. *Osteo-arthritis* of the hip is common and produces pain, difficulty in walking and limitation of movement.

THE KNEE JOINT

Type: hinge.
Bones: condyles of femur,
       upper end of tibia (Fig. 6.10),
       patella.

Special features

(a) Two semilunar cartilages, medial and lateral, of fibrocartilage, are attached

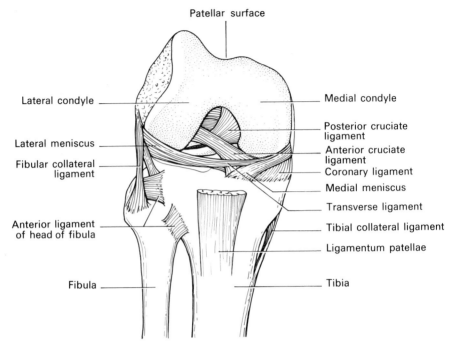

Fig. 6.9. The knee joint.

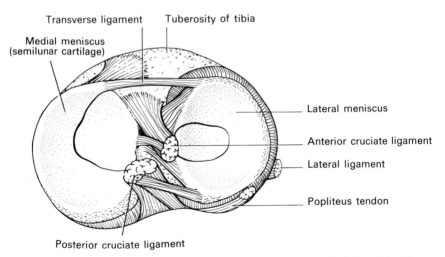

Fig. 6.10. Structures on the upper surface of the upper end of the tibia. The cruciate ligaments have been cut through.

to a rough area on the top of the tibia between the two joint surfaces; the condyles of the femur rest on them.

(b) Two strong cruciate ligaments, anterior and posterior, bind the upper end of the tibia to the intercondylar notch of the femur; they cross each other (hence the name).

(c) There are strong medial and lateral ligaments at the sides of the joint.

(d) The patella is embedded in the patellar tendon in front of the knee; it glides over the articular surface at the front of the lower end of the femur.

(e) Several bursae are present. A *bursa* is a small fluid-containing sac over which muscles and tendons move. The large suprapatellar bursa, which is continuous with the joint, lies between the lower end of the femur and the ligamentum patellae.

Main muscles: flexor—Hamstrings, gastrocnemius,
          extensor—quadriceps femoris.

Movements: flexion and extension,
          slight rotation when the knee is flexed,
          the anterior cruciate ligament prevents hyperextension,
          the posterior cruciate ligament prevents hyperflexion.

Clinical features

A *tear of a semilunar cartilage* can be the result of a sudden twist to the knee, as can occur in football or in mining with the miner in a squatting position. It is more common in the medial than the lateral cartilage because the medial cartilage is firmly attached to the capsule. Healing does not occur as the cartilage has no blood supply and after apparent recovery, attacks of giving way or locking of the knees are likely. Surgical removal is necessary. The functioning of the knee is not impaired by removal of a cartilage.

The *cruciate ligaments* may be torn in severe abduction or adduction injuries.

ANKLE JOINT

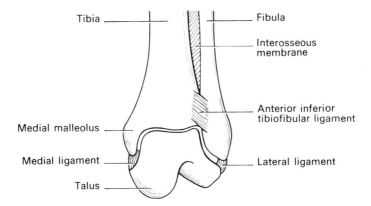

Fig. 6.11. The left ankle joint from the front.

Type: hinge joint.
Bones: lower ends of tibia and fibula,
　　　　 trochlea of the talus.
Special feature: the capsule is strengthened at the sides by strong ligaments (Fig. 6.12).

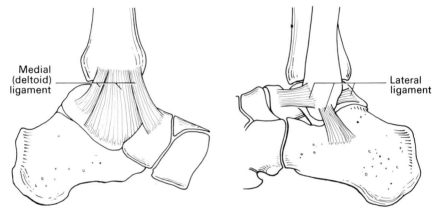

Medial (deltoid) ligament

Lateral ligament

Fig. 6.12. The ankle joint medial aspect (left), lateral aspect (right).

Main muscles: extensors of foot and toes,
　　　　　　 peroneal muscles,
　　　　　　 gastrocnemius and soleus.
Movements: dorsiflexion of foot,
　　　　　　 plantar flexion of foot.

JOINTS OF FOOT
The various tarsal bones are connected by joints at which slight gliding movements are possible.

The joints of the toes are similar to those of the fingers.

7

Muscle and Muscles

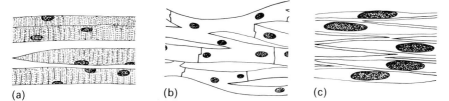

Fig. 7.1. Muscle cells: (a) striated, (b) cardiac, (c) plain.

SKELETAL MUSCLE

Skeletal muscle forms about 40 per cent of the total weight of the body. It is employed in making voluntary movements and holding the body erect.

Each muscle is enclosed within a sheath of connective tissue and is composed of many muscle fibres, which run the whole length of a muscle and are from 1 to 40 mm long. Muscle fibres are enclosed in bundles surrounded by a thin sheath. Each fibre is composed of myofibrils, each of which is composed of smaller structures called myofilaments.

Neuromotor unit

A *neuromotor unit* is composed of:
   a motor nerve cell and its axon fibre (see p.222), the muscle fibres it supplies.
   Each muscle can be considered as a large group of such units.

65

A nerve to a muscle is composed of motor and sensory fibres in about equal proportions:

A *motor nerve fibre* arises in the grey matter of the spinal cord or brainstem and supplies a group of muscle fibres, sending a branch of each fibre. Each of these branches ends on a *motor end plate* (Fig. 7.2), a structure lying against a muscle fibre.

A *sensory nerve fibre* arises in a *muscle spindle*, a long thin structure lying among the muscle fibres near the attachment of a muscle to its tendon.

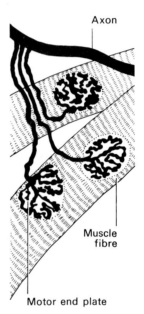

Axon

Muscle
fibre

Fig. 7.2. Motor end
plates on muscle fibres.   Motor end plate

Names of muscles

Skeletal muscles vary much in shape and size. They are called by various names, which may describe:

their position and size: e.g.
    pectoralis major — the large muscle of the chest,
    gluteus maximus — the large muscle of the gluteal region,
    latissimus dorsi — the broad muscle of the back.
their function and position: e.g.
    flexor digitorum superficialis — the superficial flexors of the fingers,
    levator ani — the raiser of the anus.
their shape and position: e.g.
    biceps brachialis — the two-headed muscle of the arm,
    rectus femoris — the straight muscle of the thigh.
the bones or cartilage they are attached to: e.g.
    sternomastoid — attached to sternum and mastoid process,
    sternothyroid — attached to sternum and thyroid cartilage.

Attachment of muscles

Most muscles are attached at both ends to bone. Some are attached to other structures—to cartilage or the skin.

The *origin* of a muscle is the more fixed point of its attachment. The *insertion* of a muscle is the usually moving part. Sometimes, in different actions, these roles are reversed.

The attachment to a bone may be:

(a) by muscle fibres running into the periosteum; the bone beneath is smooth, e.g. the front of the femur,

(b) by a mixture of muscle fibres and fibrous tissue; the bone beneath is rough, e.g. the linea aspera at the back of the femur,

(c) by a tendon; the bone may be pulled out into a tubercle or a larger projection, e.g. the biceps tubercle on the radius for the attachment of the biceps muscle.

Tendons

A *tendon* is a cord of connective tissue into which muscle can be attached and which at its other end is inserted into bone. It has some elasticity. It (a) enables a large mass of muscle to concentrate its strength on a single relatively small area of bone,

(b) enables several tendons to pass through a small space, e.g. the tendons of the muscles of the forearm as they pass in front of and behind the wrist,

(c) has a protective and supportive action around a joint.

Muscle tone

Muscle tone is the slight degree of tension in muscle at all times. It does not produce any shortening of a muscle. It is maintained by nervous discharges down the neurones of motor neurone units; only a small number of these units are in action at any one time, the particular units being repeatedly changed so that fatigue is avoided. A degree of tone is necessary to act against gravity, to maintain an erect position, and to maintain the head on the shoulders.

The sensory stimuli which stimulate the production of these motor responses come from:

the muscle spindles (the sensory organs in the muscles),

the eyes,

the vestibular organs (the organs of balance within the temporal bones).

Muscle contraction

Muscles are involved in all movements with opposing or otherwise differing actions. The following groups are involved:

prime movers: these are the muscles principally involved in making a movement;

antagonists: these are the muscles with an action the opposite to that of the prime movers. By progressively relaxing as the prime movers contract, they help to control an action and to prevent over-reaction;

fixation muscles: by increasing tension (i.e. bringing more motor neurone units into action) these fix joints that have to be fixed if an action is to be correctly carried out.

For example, in flexing the forearm on the arm
(i) the biceps and brachioradialis in the front of the arm are the prime movers,
(ii) the triceps at the back of the arm, is the antagonist,
(iii) the muscles in front of and below the shoulder are the fixation muscles which fix the shoulder so that it does not move.

THE PRINCIPAL MUSCLES OF THE BODY

Muscles of the face and scalp

The muscles of the face (Fig. 7.3) are thin, small or narrow, and are the muscles of facial expression. The *orbicularis oculi* is a ring of muscle around the eye and in the eyelids, and by contraction closes the eye. The *orbicularis oris* is a similar ring of muscle around the mouth and in the lips. Other muscles pull up or out the corners of the mouth, control the cheek (especially necessary when eating), wrinkle the skin, and wag the ears.

The *frontal muscle* is a thin flat muscle under the skin of the forehead. The *occipital muscle* is a thin flat muscle at the back of the head. They are attached at the front and back respectively of the *cranial aponeurosis*, of fibrous tissue, running over the top of the skull.

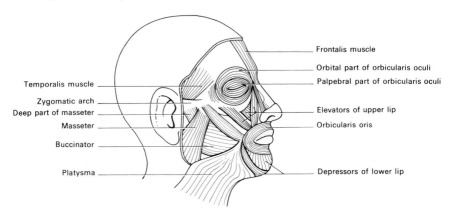

Fig. 7.3. Muscles of the face.

Muscles of mastication

The muscles of mastication open and close the mouth and move the lower jaw from side to side. The *masseter* is oblong and attached to the zygomatic arch above, and to the angle of the mandible below. The *temporalis* is a fan-shaped muscle, which arises from the side of the skull and is inserted by a tendon into the coronoid process of the mandible.

The motor branch of the trigeminal nerve (V cranial) supplies the muscles of mastication.

Muscles of the neck (Fig. 7.4)

The *sternomastoid* is the prominent muscle at the side of the neck. It is attached below to the manubrium of the sternum and the inner end of the clavicle, and, running diagonally upwards and backwards round the neck, is attached to the mastoid process of the temporal bone just behind the ear.

A group of muscles runs up the front and sides of the vertebral column and acts on the head and neck.

The upper part of the *trapezius* is situated superficially at the back of the neck.

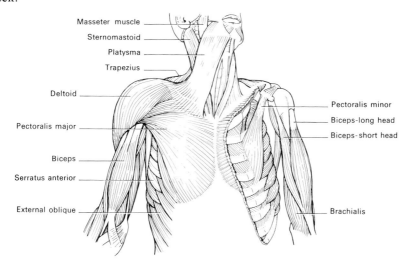

Masseter muscle
Sternomastoid
Platysma
Trapezius
Deltoid
Pectoralis major
Biceps
Serratus anterior
External oblique
Pectoralis minor
Biceps-long head
Biceps-short head
Brachialis

Fig. 7.4. Muscles of the neck and arm from the front.

Triangles of the neck. The neck is divided by the sternomastoid muscle into an anterior and a posterior triangle. The *anterior triangle* has its base at the lower border of the mandible, its anterior margin at the midline of the neck in front, and its posterior margin at the anterior border of sternomastoid. The *posterior triangle* has its base at the middle third of the clavicle, its apex at the occipital bone, its anterior margin at the posterior border of the sternomastoid, and its posterior margin at the anterior border of the trapezius.

Muscles of the arm

THE SHOULDER GIRDLE

The *trapezius* is a flat muscle which extends from the occipital bone and down the spines of the cervical and thoracic vertebrae out to the acromion and spine of the scapula. Its functions are to rotate the scapula when the arm is lifted up and control the descent of the arm, to brace the shoulder back and lift it in shrugging.

The *serratus anterior* arises by digitations from the outer surfaces of the upper eight or nine ribs and running backwards between the chest wall and the front of the scapula is inserted into the medial border of that bone. It is involved in pushing, punching and lifting the arm above the head.

Clinical feature
Paralysis of the serratus anterior causes 'winging' of the scapula, which stands away from the chest wall like an angel's wing.

The *pectoralis major* arises from the medial one-third of the clavicle, the front of the sternum and the front of the upper costal cartilages. Its fibres run outwards, twisting as they go, to be inserted into a ridge on the front of the upper end of the shaft of the humerus. The *pectoralis minor*, a small muscle lying deep to pectoralis major, arises from the front of the 2nd–5th ribs and is inserted into the coracoid process of the scapula. The functions of these muscles are to lift the arm, rotate it internally, and pull the trunk up in climbing.

Axilla (armpit)
A pyramidal space with apex pointing upwards, formed by chest wall on inner side, humerus on outer side, pectoral muscles in front, latissimus dorsi muscle behind. Contents:
    axillary artery
    axillary vein
    brachial plexus of nerves
    lymph vessels and nodes

The *deltoid* is the thick muscle over the shoulder. Its fibres arise from the lateral one-third of the clavicle and from the acromion process of the scapula, and are inserted into the deltoid tuberosity, a rough area half way down the shaft of the humerus. It is a powerful abductor of the arm, is involved in lifting the arm and controlling its descent and in both medial and lateral rotation of it.

THE UPPER ARM (Figs 7.4 & 7.5)

Anterior
The *biceps brachialis* arises by two heads: one from the coracoid process of the scapula, the other from the scapula just above the glenoid fossa. The two heads combine into one muscle, which runs down the front of the elbow, is inserted into the biceps tubercle at the upper end of the radius. The *coracobrachialis* extends from the coracoid process of the scapula to half way down the shaft of

the humerus. The *brachialis* extends from the lower half of the shaft of the humerus to the coronoid process of the ulna, lying deep to biceps and immediately in front of the elbow joint. The functions of these muscles include flexion of the elbow and supination of the forearm. It is the great strength of biceps that makes supination stronger than pronation and determines the way in which screws and corkscrews are used. Left-handed people are at a disadvantage.

Posterior
The *triceps brachialis* arises by three heads from the scapula and the back of the shaft of the humerus and running down the back of the arm is inserted into the olecranon of the ulna. It is the extensor of the elbow.

THE FOREARM

Anterior
The principal muscles (7.5) in the front of the forearm are the *superficial* and *deep flexors of the fingers*, the *flexor of the thumb*, and muscles acting on the bones at the wrist. Before entering the hand the muscles are continued as tendons. The flexors of the fingers and thumb are inserted into the phalanges. In

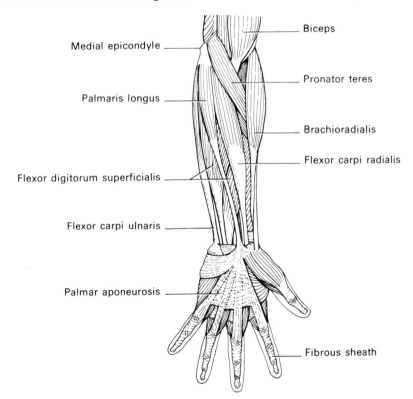

Medial epicondyle

Palmaris longus

Flexor digitorum superficialis

Flexor carpi ulnaris

Palmar aponeurosis

Biceps

Pronator teres

Brachioradialis

Flexor carpi radialis

Fibrous sheath

Fig. 7.5. Muscles of the forearm from the front.

the hand they are enclosed within synovial sheaths, of which those of the thumb and little finger extend upwards to the wrist.

Carpal tunnel
A tunnel in front of wrist, between the carpal bones behind and a ligament (the flexor retinaculum) in front. Contents:
    median nerve
    flexor tendons to fingers and thumb

Clinical features
The *carpal tunnel syndrome*—of weakness, numbness and peculiar feelings in the distribution of the median nerve to the hands and fingers—is due to pressure on the median nerve as it passes through the carpal tunnel and is relieved by cutting the flexor retinaculum. The condition is more common in women than men, because in women the tunnel is smaller.

Posterior
The extensor muscles of the wrist and fingers end in tendons which are inserted into bones at the wrist or the backs of the phalanges.

THE HAND

The *thenar muscles* are small muscles acting on the thumb and forming the thenar eminence. The *hypothenar muscles* are small muscles acting on the little finger and forming the hypothenar eminence. The *lumbricals* and *interossei* are small muscles in the palm which act on the fingers.

Muscles of the anterior and lateral abdominal wall

Muscles of the anterior and lateral abdominal wall
    rectus abdominis
    external oblique
    internal oblique
    transversus.

The *rectus abdominis* extends from the front of the costal margin above to the pubis below. It is crossed by some fibrous bands and enclosed within a sheath.

The *linea alba* is the band of fibrous tissue which, extending in the midline from the xiphoid process of the sternum to the symphysis pubis, separates the two recti abdominis muscles.

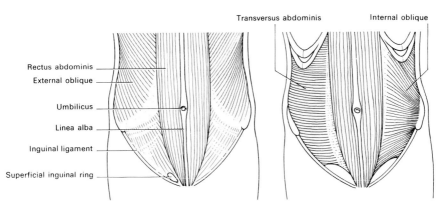

Fig. 7.6a & b. Muscles of the anterior abdominal wall.

The *external oblique*, the *internal oblique* and the *transversus* are flat muscles which form the abdominal walls at the sides and in front. The fibres of the external oblique run downwards and forwards; those of the internal oblique run upwards and forwards; those of the transversus (the deepest of the three) run transversely. In front the three muscles end in a common sheath which encloses the rectus abdominis.

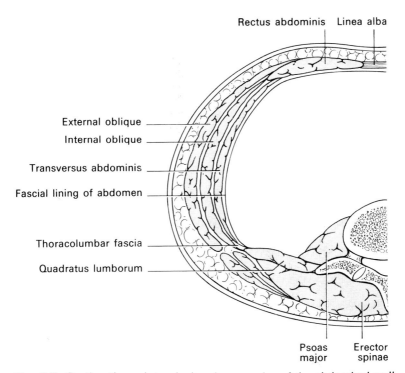

Fig. 7.7. Section through trunk showing muscles of the abdominal wall.

The *inguinal ligament* (Poupart's) is a ligament which runs downwards and forwards from the anterior superior iliac spine to the pubic bone.

Inguinal canal
A musculofibrous canal about 3.5 cm long in the lower part of anterior abdominal wall just above inguinal ligament. Contents:
  in men
  spermatic cord
  testicular blood vessels
  and lymph vessels
  some nerves
  in women
  round ligament
  lymph vessels

Clinical features

In operating through the anterior abdominal wall the surgeon endeavours as much as he can to divide the muscles in the direction of their fibres so as to inflict as little damage as possible upon them. An *inguinal hernia* is a hernia (rupture) of peritoneum or bowel out of the abdomen and into the inguinal canal. The upper opening of the inguinal canal is one of the weak spots in the abdominal wall. An *undescended testis* (one which has not descended to its normal place in the scrotum) is sometimes found in the inguinal canal.

Muscles of the posterior abdominal wall

The *psoas* arises from the sides of the bodies of the lumbar vertebrae; its fibres pass downwards and are joined by fibres of *iliacus*, which arises from the innominate bone. Passing into the thigh in front of the hip joint, the fibres of these two muscles are inserted by a common tendon into the lesser trochanter of the femur. The action of these muscles is to flex the thigh upon the pelvis, to rotate the thigh medially and to help to maintain the upright position.

The *quadratus lumborum* is a short square muscle at the back of the abdomen, extending from the 12th rib above to the iliac crest below. It keeps the rib steady in respiration so that the diaphragm can pull upon it.

The diaphragm

The diaphragm separates the thorax from the abdomen. Its fibres (Fig. 7.8) arise from:
  the xiphoid cartilage, in front
  the inner surfaces of the lower ribs, at the sides
  the bodies of the lumbar vertebrae, behind, by two muscular bands called crura.

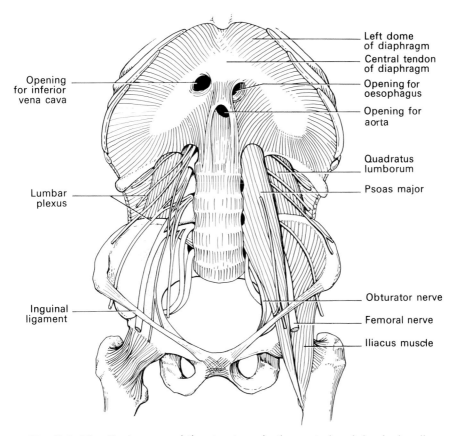

Fig. 7.8. The diaphragm and the structures in the posterior abdominal wall.

From all these sources fibres run upwards and centrally to be inserted into a broad trefoil-shaped tendon.

> *Structures passing through the diaphragm*
> aorta: between the crura at the back
> thoracic duct: with the aorta
> inferior vena cava: through the tendon
> oesophagus: posteriorily and to the left

Nerve supply: mainly by the two phrenic nerves, which arise from C3, 4 and 5.

The diaphragm is the principal muscle of respiration. Contraction of its fibres causes the central tendon to be pulled downwards and the space within the chest to be enlarged.

Clinical features

The diaphragm is formed by muscle from several sources and weak places may develop at the places where various muscles should have joined. A *diaphragmatic hernia*, of peritoneum or stomach, can occur at one of these places, causing indigestion, epigastric discomfort and other symptoms.

Muscles of the pelvis

These muscles form the floor of the pelvic cavity.

The *levator ani* is a cone-shaped muscle which arises from the pubic bone in front, from the tuberosity of the ischium behind, and by a fibrous band

---

Perineum
The connective tissue and small muscles at the outlet of pelvis below. It is traversed by:
   urethra
   vagina in women
   anus

---

stretched between these two attachments. The fibres run downwards and inwards to blend with the sphincter muscles of the anus. It supports the pelvic

---

Ischiorectal fossae
Wedge-shaped spaces, one on each side of the anus and lower end of the rectum. Base the perineum, outer wall the pelvis, inner wall levator ani muscle. Contents:
   fat

---

viscera and contracts during defecation, pulling the anus upwards. Its anterior fibres in men support the prostate gland and in women form a sphincter round the vagina.

Muscles of the back (Fig. 7.9)

The *erector spinae* consists of a mass of muscle fibres arising from the back of the sacrum and adjacent part of the innominate bone and being attached to the back of the vertebral column above, with further fibres arising from the back of the vertebrae and up to the occipital bone of the skull. It maintains the erect position of the trunk and enables the trunk to regain that position when it has been flexed.

The *latissimus dorsi* is a broad flat muscle at the back of the chest. It arises from the spines of the lower six thoracic vertebrae and adjacent structures and running upwards and outwards is inserted into the upper end of the shaft of the humerus, forming the posterior wall of the axilla. Its main actions are pulling the arm down against resistance, rotating the arm inwards, and pulling the trunk towards the arms in climbing. In forced respiration it compresses the posterior part of the abdomen.

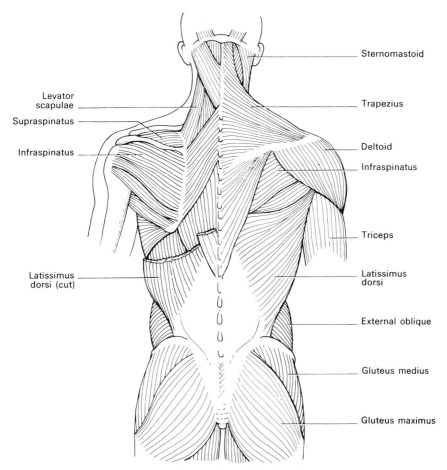

Fig. 7.9. Muscles of the back.

Muscles of the leg

BUTTOCK

The *gluteus maximus, gluteus medius* and *gluteus minimus* are the muscles of the buttock. They all arise from the outer surface of the ilium, with the gluteus maximus also arising partly from the back of the sacrum. Passing outwards they are inserted into the greater trochanter and upper part of the shaft of the femur. Their principal actions are in maintaining the erect position, extending the hip joint in running, climbing, and going upstairs, in raising the trunk from the sitting or stooping position, in laterally rotating and abducting the thigh.

THIGH

Anterior group (Figs 7.10, 7.11)
The *quadriceps femoris* is composed of four muscles: *rectus femoris, vastus*

medialis, vastus lateralis, vastus intermedius. They arise from the innominate bone and from the shaft of the femur and run together to be inserted into the *patellar tendon*, which passing downwards in front of the knee is inserted into a tubercle on the front of the upper end of the tibia. The patella is enclosed within this tendon. The *sartorius* is a long narrow muscle which runs obliquely downwards and across the quadriceps from the anterior superior iliac spine to the inner side of the tuberosity of the tibia. The principal actions of this group are to flex the hip and to extend the knee and to assist in maintaining the stability of the knee and the erect posture.

Medial group
The *adductor longus, adductor brevis* and *adductor magnus* are on the inner side of the thigh. They arise on the pubic bone and running downwards and outwards are inserted into the linea aspera on the back of the shaft of the femur. They adduct the thigh and assist in maintaining the stability of the hip joint.

Femoral triangle
A slightly hollowed-out triangle in front of thigh, formed by inguinal ligament above, the adductor longus muscle medially, and the sartorius muscle laterally, with a floor formed by other muscles.
Contents (from without inwards):
  femoral nerve
  femoral artery
  femoral vein
  femoral canal (about 1.25 cm long, containing
    connective tissue and a lymph node)

Posterior group
The *biceps femoris, semimembranosus* and *semitendinosus* are the hamstrings at the back of the thigh. They arise from the ischial tuberosity. They run downwards. Just above the back of the knee the biceps passes laterally to be inserted into the head of the fibula and the other two pass medially to be inserted into the upper end of the tibia. Their principal action is flexion of the knee.

Clinical features
A *femoral hernia* is a hernia into the femoral canal, whose upper end is one of the weak spots of the abdominal wall. It is more common in women than men because in women the pelvis is wider and the canal correspondingly wider. The hernia lies to the medial side of the femoral vein and lateral to a small ligament.

LEG BELOW KNEE

Anterior group (Figs 7.10, 7.11)
The *tibialis anterior, extensor hallucis longus* and *extensor digitorum longus* arise

from the tibia, fibula and the interosseous ligament between the two bones, run down the leg, become tendons (which are fastened down at the ankle by bands of fascia) and are inserted through them into bones of the tarsus and phalanges. Their principal action is dorsiflexion of the foot and toes.

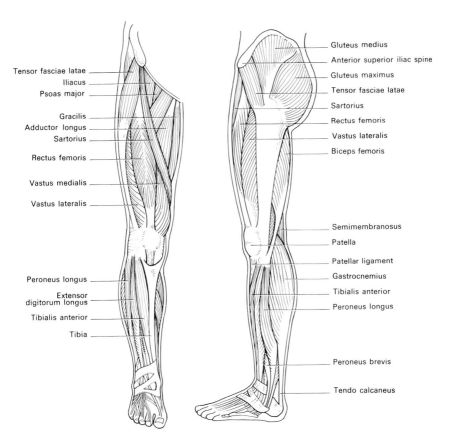

Figs 7.10 & 7.11. Muscles of the leg, anterior and lateral aspects.

Posterior group (Figs 7.12, 7.13)

The *gastrocnemius, soleus, tibialis posterior* and *flexor digitorum longus* form the calf. The gastrocnemius arises by two tendons, one from each condyle of the femur. As its fibres run downwards they are joined by the fibres of soleus and the two muscles run into the *tendo Achilles*, which is inserted into the back of the calcaneum. Tendons from the other muscles run behind the medial malleolus to enter the sole and be inserted into bones of the tarsus and into phalanges. Gastrocnemius and soleus are powerful plantar flexors of the foot, help to maintain the balance, and are the principal force in walking, running and jumping. Gastrocnemius is also a flexor and stabilizer of the knee.

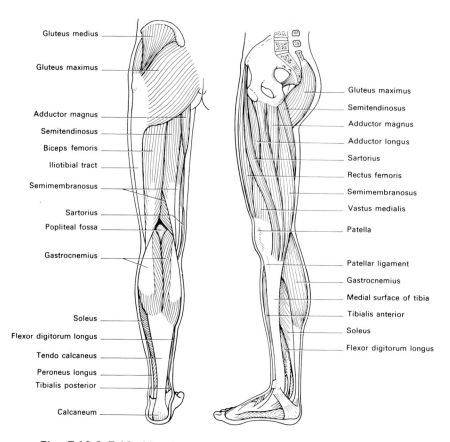

Figs 7.12 & 7.13. Muscles of the leg, posterior and medial aspects.

Popliteal fossa
A diamond-shaped space behind the knee; bounded above by biceps laterally and semimembranosus and semitendinosus medially and below by the heads of gastrocnemius. Floor formed by lower end of femur and capsule of knee joint. Contents:
  popliteal artery
  popliteal vein
  medial popliteal nerve
  lymph nodes
  fat

Peroneal group

The *peroneus longus* and *peroneus brevis* arise from the fibula, run down the lateral side of the leg, and pass behind the external malleolus to be inserted into tarsal and metatarsal bones. Their action is to evert the foot and assist in plantar flexion.

THE FOOT

In addition to the tendons of the flexor muscles passing through it, the foot contains several small muscles which act on the toes.

The *foot* is an arched structure with longitudinal and transverse arches. The weight of the foot is borne mainly on the heads of metacarpals and the calcaneum.

> *Arches of foot*
> The arches are maintained by'
>   shapes of tarsal bones
>   strong ligaments between the bones
>   short muscles of foot
>   tendons of flexor and peroneal muscles
>   plantar aponeurosis—a tough, thick layer of
>   fibrous tissue in the sole

Clinical feature

Flat footedness is usually due to a minor deformity of the tarsal bones with loss of the normal longitudinal arch. Treatment is by strengthening the foot muscles by exercises or by supporting the arch with a rigid support.

FASCIA

A fascia is a sheet of fibrous tissue, with or without fat, separating one part of the body from another.

Subcutaneous fascia is a complete subcutaneous sheet of fibrous and fatty tissue. As fat is a poor conductor of heat, the subcutaneous fat helps to keep the body warm.

Deep fascia is tough fibrous fat-less tissue which (a) surrounds the muscles with a fascial sheath, (b) provides additional attachments for some muscles, (c) enables structures to glide over one another, and (d) provides beds in which lie blood vessels and nerves.

Clinical features

The position and attachments of the various fascial beds determine the direction in which blood, pus and other fluids can move among the tissues and organs. It is by the movement of fascia away from trouble that structures are pulled out of the route of penetrating weapons, which might be thought to have transfixed them.

8
The Cardiovascular System. Part I: the Heart

The cardiovascular system:
  heart
  arteries and arterioles
  thoroughfare vessels
  capillaries and sinusoids
  venules and veins

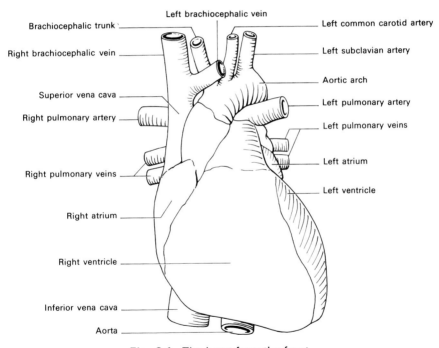

Fig. 8.1. The heart from the front.

The heart (Fig. 8.1) is about the size of one's clenched fist and is situated within the chest, its right border being just to the right of the sternum and its apex in the fifth left intercostal space in the mid-clavicular line (Fig. 8.2).

   Its relations are:
above: the great vessels (the aorta, pulmonary trunk, etc),

83

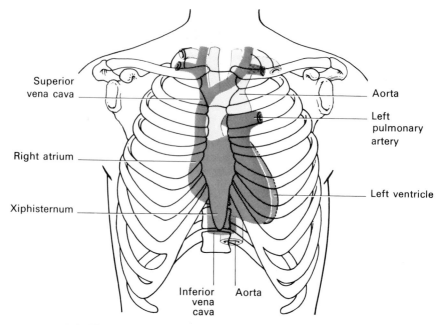

Fig. 8.2. The position of the heart and great vessels within the chest.

below: the diaphragm,
on either side: the lungs,
behind: the descending aorta, oesophagus and spinal column.

Chambers of heart
right atrium
right ventricle
left atrium
left ventricle

Right atrium

The right atrium (Fig. 8.3) is on the right side of the heart and lies mostly behind the sternum. Blood enters it through:

the superior vena cava at its upper end,

the inferior vena cava at its lower end,

the coronary sinus (a small vein through which comes blood from the heart itself).

The *right auricle* is a small pointed projection from the atrium, lying in front of the origins of the aorta and pulmonary artery.

On the left side of the atrium the *right atrioventricular opening* opens into the right ventricle.

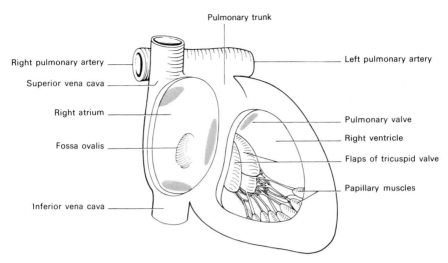

Fig. 8.3. The right side of the heart.

Right ventricle

The right ventricle (Fig. 8.3) is a thick-walled chamber which forms much of the front of the heart.

The *right atrioventricular (tricuspid) valve* guards, on the ventricular side, the right atrioventricular opening. This valve, like the other heart valves, is composed of a thin sheet of fibrous tissue covered on each side by endocardium. The tricuspid valve is composed of three flaps. The base of each flap is attached to the edge of the opening. The free border of each flap is attached by *chordae tendineae* (thin strands of fibrous tissue) to small conical projections of muscle tissue arising from myocardium and projecting into the ventricle.

The *pulmonary opening* into the pulmonary artery is at the upper end of the ventricle and is guarded by a *pulmonary valve*, composed of three semilunar flaps.

Left atrium

The left atrium is a thin-walled cavity situated at the back of the heart. Two pulmonary veins enter it on each side, bringing blood from the lungs. The atrium opens below into the left ventricle through the left atrioventricular opening.

The *left auricle* is a small pointed projection from the atrium, lying to the left of the origin of the aorta.

Left ventricle

The left ventricle (Fig. 8.4) is a thick-walled cavity at the left and back of the heart. Its wall is about three times as thick as the wall of the right ventricle. The

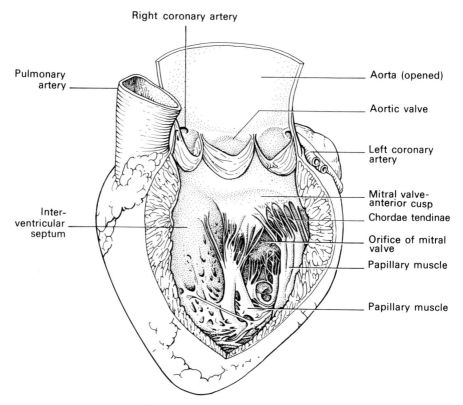

Right coronary artery

Pulmonary
artery

Aorta (opened)

Aortic valve

Left coronary
artery

Mitral valve-
anterior cusp

Chordae tendinae

Inter-
ventricular
septum

Orifice of mitral
valve

Papillary muscle

Papillary muscle

Fig. 8.4. The left ventricle and aortic valve.

left atrioventricular (*mitral*) *valve* surrounds the left atrioventricular opening on
the ventricular side; it has two flaps (getting its name from a resemblance to a
bishop's mitre), the borders of which are attached to chordae tendineae, which

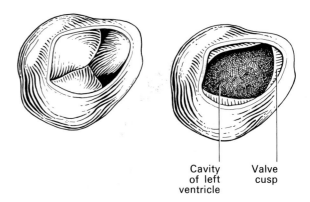

Fig. 8.5. The aortic
valve, closed (left) and
open (right). The
pulmonary valve has the
same appearance.

Cavity
of left
ventricle

Valve
cusp

are attached to conical projections of myocardium on the ventricular wall. The *aortic opening* opens from the upper end of the ventricle into the aorta and is guarded by an aortic valve (Fig. 8.5) of three cusps, similar to those of the pulmonary valve.

> *Tissues of heart*
> myocardium
> endocardium
> pericardium

Myocardium

Myocardium forms the greatest part of the wall of the heart. It is composed of cardiac muscle fibres, which are striated and connected to one another by muscular branches. The fibres begin to contract in the embryo before any nerves reach them, and continue to contract rhythmically even if deprived of any nerve supply.

Endocardium

The endocardium lines the inside of the chambers of the heart and covers the valves on both sides. It is composed of a layer of endothelial cells, under which are layers of connective tissue; it is smooth and shining.

Pericardium

The pericardium is the fibrous bag in which the rest of the heart is enclosed. It is a double layered sac: the two layers are in contact with and glide over one another with the aid of the fluid which they secrete and which moistens their surfaces. The amount of fluid normally present is about 20 ml. At the root of the heart (where the great vessels, lymphatics and nerves enter it) the two layers are continuous. There is a layer of fat between the myocardium and the layer of pericardium over it.

The coronary arteries

The two coronary arteries (Fig. 8.6), right and left, supply the wall of the heart with blood. They arise from the aorta immediately above the aortic valve and run down the surface of the right and left sides of the heart respectively, giving off deep branches to the myocardium. They supply their respective sides of the heart; but there are individual variations, and in some people the right coronary artery supplies part of the left ventricle. There are relatively few anastomoses between right and left artery.

Clinical feature

Degeneration of the arterial wall can spread from the aorta into the coronary arteries, reducing the supply of blood to the heart. *Angina pectoris* is a painful condition of the chest, left arm, and surrounding area due to a diminished supply of blood to the heart. A *coronary thrombosis* is a clotting of blood in a

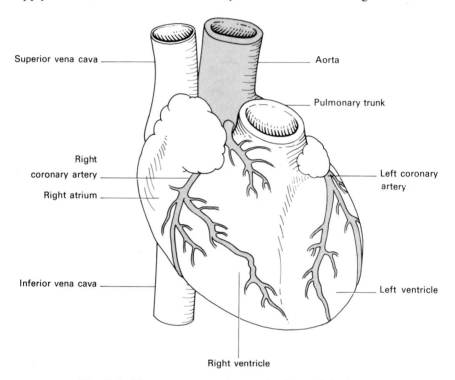

Fig. 8.6. The coronary arteries on the front of the heart.

degenerating coronary artery. Owing to the paucity of interarterial connections, the function of one artery cannot be taken over by the other, and blocking of one coronary artery can produce sudden death or (if the patient survives) severe damage to the myocardium and a reduction in cardiac efficiency.

The cardiac cycle

The cardiac cycle is the sequence of events during one beat of the heart. It occurs in two phases: diastole and systole.

DIASTOLE

Diastole is the period of rest which follows the period of contraction.

At first:

1. Venous blood is entering the right atrium through the superior and inferior venae cavae.

2. Oxygenated blood is passing into the left atrium through the pulmonary veins.

3. The two atrioventricular (AV) valves (tricuspid and mitral) are closed and blood is prevented from passing from atrium into ventricle.

4. The pulmonary and aortic valves are closed, preventing the return of blood from the pulmonary artery into the right ventricle and from the aorta into the left ventricle.

Then:

5. As more blood enters the atria the pressure in them rises; and when the pressure in them is greater than in the ventricles, the AV valves open and blood begins to flow from the atria into the ventricles.

SYSTOLE

Systole is the period of muscular contraction. It lasts for 0.3 seconds.

1. Stimulated by the sino-atrial node, the walls of the atria contract, driving the rest of the blood from them into the ventricles.

2. The ventricles expand to take in this blood from the ventricles and then start to contract.

3. As the pressure in the ventricles becomes greater than that in the atria, the AV valves close. The chordae tendineae prevent their being pushed into the atria.

4. The ventricles continue to contract. The pulmonary and aortic valves open under this increasing pressure.

5. Blood is expelled from the right ventricle into the pulmonary artery and blood from the left ventricle is expelled into the aorta.

6. Muscular contraction then stops, and with the beginning of muscular relaxation a new cycle begins.

Each contraction is followed by a short absolute refractory period in which no stimulus can produce a contraction, and this by a short relative refractory period in which contraction requires a strong stimulus.

The heart beat

The *sino-atrial node* (SA node or pacemaker of the heart) is a small area of muscle fibres and nerve cells situated in the wall of the heart close to the entry of the superior vena cava. At the beginning of systole, a wave of contraction starts in this node and:

(a) spreads through the walls of both atria, simulating them to contract; this atrial contraction does not spread to the ventricles because it cannot get through a ring of fibrous tissue which separates the atria from the ventricles,

(b) reaches and stimulates the atrioventricular node.

The *atrioventricular node* (AV node) is a small area of specialized tissue in the wall between the right atrium and ventricle.

The *atrioventricular bundle* (bundle of His) is a band of muscle and nerve fibres which runs in the septum between the two ventricles, reaches the apex of the heart, and there divides into two main branches, one for each ventricle, which divide into smaller and smaller branches in the wall of the ventricles (Fig. 8.7).

A wave of contraction spreads from the AV node down the AV bundle and sets off the contraction of the two ventricles simultaneously. The wave of contraction which began in the SA node causes the atria to contract just before the ventricle because the wave reaches the atria immediately and the one to the ventricles has to go down the AV bundle.

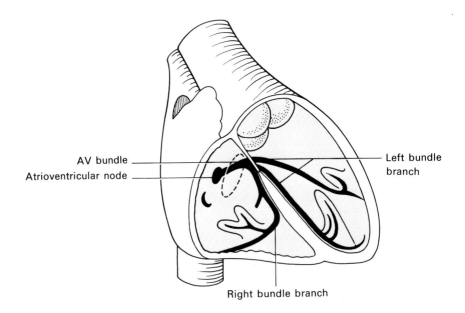

Fig. 8.7. The atrioventricular bundle and its branches.

Nervous control of the heart

Although the heart is capable of beating on its own and adjusting the rate and force of its beat to the amount of blood entering it, it has a double nerve supply, which controls the functioning of the SA node and prepares the heart for changing conditions.

Sympathetic fibres pass from the ganglia on the cervical part of the sympathetic trunk and transmit impulses which stimulate the SA node into faster activity and increase the force of contraction.

Parasympathetic fibres reach the heart through branches of the vagus (X cranial) nerve and transmit impulses which slow down the SA node and reduce the force of contraction.

The *higher nerve centres* involved are:

cerebral cortex,

hypothalamus,

a cardiac centre in the medulla oblongata, composed of:

(a) a cardio-accelerator centre,

(b) a cardio-inhibitor centre.

Cardiac output

The output of blood from the heart depends upon:

the *heart rate*: at rest usually about 70 beats a minute,

the *stroke volume*: the amount of blood discharged from a ventricle at each beat. At rest this is usually about 70 ml. In mild exercise it is increased to 125 ml.

At the beginning of its contraction a ventricle, with the body at rest, contains about 120 ml. About 50 ml is left in the ventricle at each beat.

The total discharge per minute is about 5 litres.

The *heart rate* is controlled:

(a) mainly by a reduction in the stimulation through the parasympathetic (vagal) nerve fibres,

(b) to a lesser extent by stimulation through sympathetic nerves.

The *stroke volume* is controlled by changes in the length of the cardiac muscle fibres. The greater the length (in healthy muscle) the greater the contraction. When more blood enters the heart (as in exercise) the greater is the contraction and therefore the greater the stroke volume.

The cardiac output is measured by: (a) measuring the amount of oxygen taken up by the lungs per minute, (b) various dilution techniques involving dyes, radio-active isotopes etc.

Heart failure

Heart failure occurs when the output of the heart is insufficient to supply the body's metabolic requirements. Normally the heart can without trouble increase its output several times when, as in exercise, the metabolic requirements of the body are increased. A mild defect of cardiac function will produce evidence of heart failure during exercise. With a progressively greater defect, symptoms appear with less and less effort and in severe failure are present at rest.

Acute heart failure occurs when a clotting of blood in a coronary artery or a pulmonary artery suddenly reduces the heart's efficiency. Certain compensatory mechanisms come into action, such as an improvement of cardiac contraction, a better return of blood to the heart, the direction of blood away from organs of relatively less important to the two vitally important ones, the heart and the brain.

Causes of chronic heart failure
  valvular disease of heart
  disease of heart muscle
  fatty degeneration of heart
  hypertension (high blood pressure)

Clinical features vary with failure of the right and the left ventricle. They include:

fatigue,

dyspnoea (shortness of breath),

orthopnoea (an inability to breathe lying down),

cough and haemoptysis (coughing up of blood),

enlargement of the liver,

oedema (excessive tissue fluid) in ankles etc.

Electrocardiograph

An electrocardiograph (ECG) (Fig. 8.8), is a recording of the electrical changes that take place in the heart as a result of its beat.

A normal ECG shows:

P wave: produced by contraction of the atria; last for 0.10 second.

QRS complex: produced by contraction of the ventricles; lasts for up to 0.09 second.

T wave: produced by ventricular relaxation.

PR interval: is the time taken for the impulse to pass down the ventricular bundle.

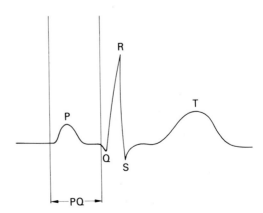

Fig. 8.8. The waves of an electrocardiograph.

The ECG is recorded on paper running at a speed of 25 mm per second, each millimetre on the paper representing 0.04 sec; the heavier lines marked on the paper every 5 mm represent 0.2 sec. The ECG is recorded twelve times in each patient, using twelve different electrode positions. The electrical activity of the heart is the same, but the waves differ as the leads pick it up from different positions.

Abnormalities of the size, shape and position of the waves and of the times involved occur in many cardiac diseases including ventricular hypertrophy, any blocking of the AV bundle and its main branches, infarction of the heart by blocking of a coronary artery. Abnormal cardiac rhythms can be measured.

In normal people at any age the ECG (taken continuously for 24 hours) can show short disturbances, such as an occasional extra beat, coupling of two beats together or a burst of rapid irregular activity.

The heart sounds

The heart produces sounds during its beating, sounds which can be heard if the ear is placed on the chest wall or with the aid of a stethoscope.

1st heart sound
Sounds like 'lub' spoken softly. It is produced by the sudden tensing of the mitral and tricuspid valves at the onset of ventricular systole.

Splitting of the 1st heart sound into two is due to asynchronous closing of the two valves as a result of one centricle contracting slightly after the other.

2nd heart sound
Sounds like 'dub'. It is produced by the vibrations caused by closure of the aortic and pulmonary valves.

Splitting of the 2nd heart sound into two during inspection is normal and best heard in young people. It is due to a slight delay in the closure of the pulmonary valve because of the flow of blood into the right ventricle.

Other sounds which may be heard are:

3rd heart sound
This is a low soft thud heard after the second sound in most children and some young adults. It is due to a sudden tightening of the mitral valve cusps.

4th heart sound
A low soft sound preceding the 1st heart sound and heard when either atrium contracts with more force than the other.

The diaphragm of a stethoscope is used for listening to high frequency sounds. The bell is used for listening to low frequency sounds.

Clinical features
Murmurs are rushing noises heard when there is a turbulence in the blood as it flows through the heart. The normal flow of blood is inaudible.

Causes of heart murmurs
  leakage of blood backwards through a faulty
    valve
  obstruction to flow of blood by a stiff, deformed
    valve
  congenital lesions of heart
  high cardiac output (as in anaemia and
    hyperthyroidism) causing 'flow murmurs'

Phonocardiography is a method of picking up heart sounds and murmurs by microphones applied to the chest and connected to a recording apparatus.

9
The Cardiovascular System. Part II: the Blood Vessels

Blood vessels
  arteries
  arterioles
  thoroughfare vessels
  capillaries
  sinusoids
  venules
  veins

THE ARTERIES

The arteries are the tubes through which blood is transmitted to the tissues and organs. They are composed of:
  an inner layer or *intima*: the smooth inner lining,
  a *middle coat* of *elastic tissue* or *muscle*: the aorta and its large branches have a middle coat composed of elastic tissue (as their function is that of conducting blood to the organs); the smaller arteries have a middle coat of muscle (which can regulate the amount of blood supplied to an organ); the change from one type of tissue to the other is gradual,
  an *outer layer* of *connective tissue*.

The aorta

The aorta (Fig. 9.1) is the main artery of the body. It consists of the thoracic aorta within the chest and its continuation and abdominal aorta within the abdomen.

The *thoracic aorta* begins at the aortic orifice of the left ventricle. It consists of three parts:
(a) *ascending aorta*: is about 5 cm long and passes upwards and to the right,
(b) *arch of the aorta*: arches backwards and to the left over the left bronchus to reach the left side of the 4th thoracic vertebra (Fig. 9.2),
(c) *descending aorta* passes downwards at the back of the thorax, between the heart and the thoracic part of the spinal column; it passes through the aortic opening of the diaphragm to become the abdominal aorta.

The *abdominal aorta* begins at the aortic opening and passes downwards in front of the lumbar part of the vertebral column to end in front of the body of the 4th lumbar vertebra by dividing into right and left common iliac arteries.

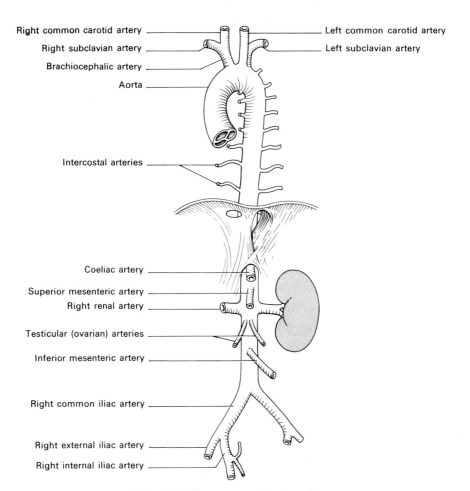

Right common carotid artery

Right subclavian artery

Brachiocephalic artery

Aorta

Left common carotid artery

Left subclavian artery

Intercostal arteries

Coeliac artery

Superior mesenteric artery

Right renal artery

Testicular (ovarian) arteries

Inferior mesenteric artery

Right common iliac artery

Right external iliac artery

Right internal iliac artery

Fig. 9.1. The aorta and its branches.

Branches of aorta
Within the thorax
    coronary arteries, right and left
    brachiocephalic artery
    left common carotid artery
    left subclavian artery
Within the abdomen
    coeliac artery
    renal arteries, right and left
    testicular arteries, right and left (in male)
    ovarian arteries, right and left (in female)
    superior mesenteric artery
    inferior mesenteric artery
    lumbar arteries, right and left
    common iliac arteries, right and left

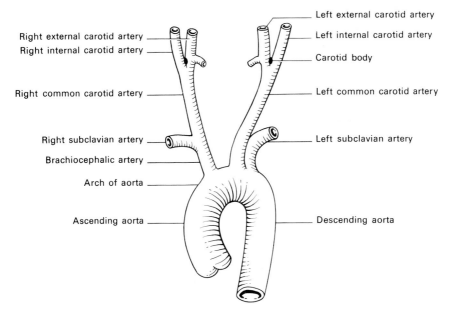

Fig. 9.2. The arch of the aorta and its branches.

Arteries to organs and structures

HEART

The heart is supplied by the *right* and *left coronary arteries* (see p. 87).

HEAD AND NECK

The head and neck are supplied by the *common carotid arteries*. There is a difference between the right and left sides as the right common carotid artery is a branch of the short brachiocephalic artery and the left common carotid artery arises directly from the aorta.

On each side the common carotid artery runs up the neck under cover of the sternomastoid muscle and divides at the level of the upper border of the thyroid cartilage into the external and internal carotid artery.

The *external carotid artery* supplies the neck and head through its branches, the largest of which are:

superior thyroid artery: to the thyroid gland,
lingual artery: to the tongue,
facial artery: to the face,
occipital artery: to the back of the head,
superficial temporal artery: to the front and side of the head,
maxillary artery: to structures at the back of the upper jaw.

The *middle meningeal artery* (one of the branches of the maxillary artery) passes through a small hole in the skull to supply the meninges. It can be torn in a fracture of the skull.

BRAIN

The arteries supplying the brain are:

the right and left internal carotid arteries,
the right and left vertebral arteries.

The *internal carotid artery* on each side is one of the two terminal branches of the common carotid artery. It runs deep in the neck and passes through the carotid canal in the temporal bone of the skull to emerge within the skull, where it runs forwards, lying in the cavernous sinus (a venous sinus on the lateral side of the body of the sphenoid bone), and ends by dividing into the anterior and middle cerebral arteries.

The *vertebral artery* is a branch of the subclavian artery at the root of the neck. It runs upwards through the foramina in the transverse processes of the upper six cervical vertebrae and then through the foramen magnum into the skull. At the junction of the pons and the medulla oblongata of the brain, the two vertebral arteries join to form the *basilar artery*, which gives branches to the cerebellum, medulla oblongata and pons, and ends by dividing into the right and left posterior cerebral arteries.

The *arterial circle*, also called the circle of Willis (Fig. 9.3), is a ring of arteries at the base of the brain formed by:
(a) the two anterior cerebral arteries and a short anterior communicating artery connecting them,
(b) the middle cerebral artery on each side,
(c) the posterior communicating artery, a branch connecting the middle and posterior cerebral artery on each side,
(d) the posterior cerebral artery on each side.

The communicating arteries can vary in size or be absent. Normally the communications between these arteries is so good that blocking of one of them does not impair the blood supply of the brain.

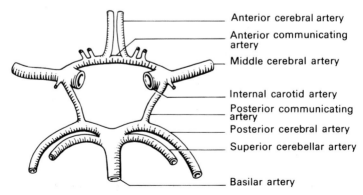

Anterior cerebral artery
Anterior communicating artery
Middle cerebral artery
Internal carotid artery
Posterior communicating artery
Posterior cerebral artery
Superior cerebellar artery
Basilar artery

Fig. 9.3. The arterial circle. The communicating arteries may not always be present.

The *anterior cerebral artery* (Fig. 9.4) on each side runs along the top of the corpus callosum (the thick band of fibres connecting the two cerebral hemispheres) and supplies the frontal and parietal lobes. This and the other cerebral arteries give off perforating branches to supply structures inside the brain, but the anastomoses between these and the superficial arteries are usually not adequate to maintain the blood supply should one of them be blocked.

The *middle cerebral artery* (Fig. 9.4) runs in the lateral sulcus of the brain

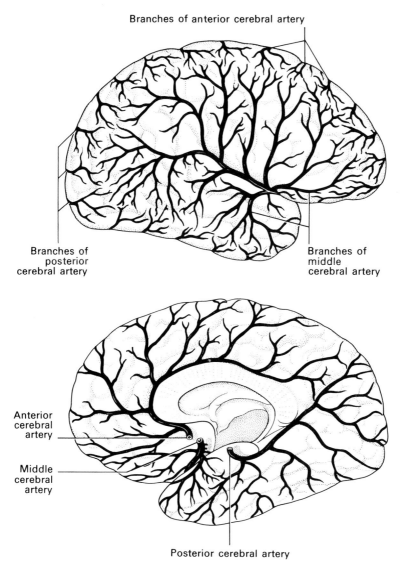

Branches of anterior cerebral artery

Branches of posterior cerebral artery

Branches of middle cerebral artery

Anterior cerebral artery

Middle cerebral artery

Posterior cerebral artery

Fig. 9.4. The blood supply to a cerebral hemisphere by the anterior, middle and posterior cerebral arteries. Lateral aspect above, medial aspect below.

and supplies parts of the frontal, parietal and occipital lobes. On the left it is the
artery to the speech centre of the brain. One of its perforating branches is called
the artery of cerebral haemorrhage as it is liable to rupture in old age.

The *posterior cerebral artery* (Fig. 9.4) supplies the occipital lobe and part of
the parietal lobes. It is the artery to the visual area of the brain.

THE EYE

The eye is supplied by the *ophthalmic artery*, a branch of the internal carotid
artery.

THE ARM

The arm (Fig. 9.5) is supplied by the subclavian artery and its branches.

The *right subclavian artery* is a branch of the short brachiocephalic artery;
the *left subclavian artery* of the arch of the aorta. It runs deep in the bottom of
the neck, then between the clavicle and the first rib. It is continued in the axilla
as the *axillary artery*. The axillary artery is continued as the *brachial artery*,
which runs down the inner side of the arm under cover of the biceps muscles and
then, coming to the front of the elbow, divides into the radial and the ulnar
arteries. The *radial artery* runs down the front of the outer side of the forearm
and under muscular cover for most of the way; it is palpable as the pulse at the
wrist. The *ulnar artery* runs down the inner side of the front of the forearm
under muscular cover. The radial and ulnar arteries are connected in the palm
by two palmar arches. All these arteries give off branches to the skin, muscles
and joints of the arm.

THE BREAST

The breast is supplied by (a) branches coming through the intercostal spaces
from the internal mammary artery, which runs down the inside of the chest
behind the costal cartilages, and (b) branches from the axillary artery.

Branches of the abdominal aorta (Fig. 9.6)

Branches of abdominal aorta
coeliac artery
renal arteries, right and left
testicular arteries, right and left (in male)
ovarian arteries, right and left (in female)
superior mesenteric artery
inferior mesenteric artery
lumbar arteries, right and left
common iliac arteries, right and left (terminal
branches)

The *coeliac artery* is a short artery which runs forwards from the aorta and divides into:

the *splenic artery* which supplies the spleen and stomach,

the *hepatic artery* to the liver; it gives off branches to the stomach and gall-bladder,

the *left gastric artery* to the stomach.

The *renal arteries* are large arteries running transversely to the kidneys.

The *testicular arteries* (in the male) arise just below the renal arteries and run

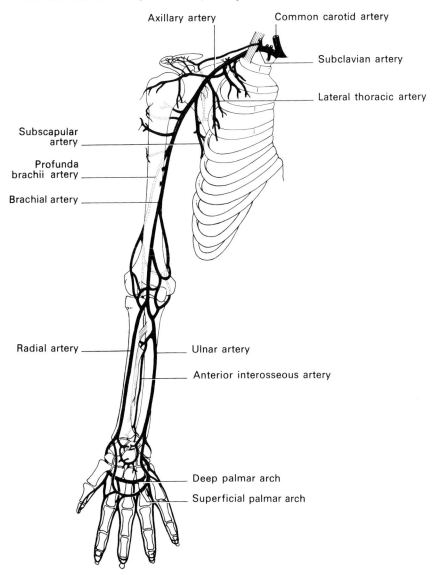

Fig. 9.5. The blood supply to the arm.

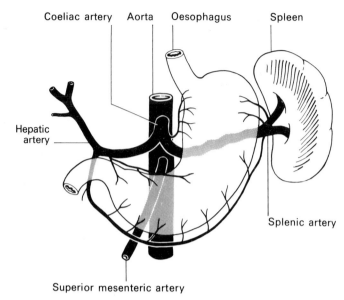

Coeliac artery   Aorta   Oesophagus   Spleen

Hepatic
artery

Splenic artery

Superior mesenteric artery

Fig. 9.6. The blood supply to the stomach and spleen.

down behind the abdominal contents, enter the spermatic cord and supply the testis and epididymis.

The *superior mesenteric artery* (Fig. 9.7) arises from the front of the aorta and runs downwards in the mesentery to divide into a large number of branches which supply the small intestine and the large intestine as far as the middle of the transverse colon.

The *inferior mesenteric artery* (Fig. 9.8) arises on the left side of the aorta and runs downwards and to the left to supply the large intestine from the middle of the transverse colon to the middle of the rectum.

Arteries to abdominal and pelvic organs
   stomach: splenic artery, hepatic artery, left
     gastric artery
   liver: hepatic artery
   spleen: splenic artery
   pancreas: coeliac artery, superior mesenteric
     artery
   small intestine: superior mesenteric artery
   large intestine: superior and inferior mesenteric
     arteries
   rectum: inferior mesenteric artery, internal iliac
     artery
   testis and epididymis: testicular artery
   ovary: ovarian artery
   uterus: uterine artery

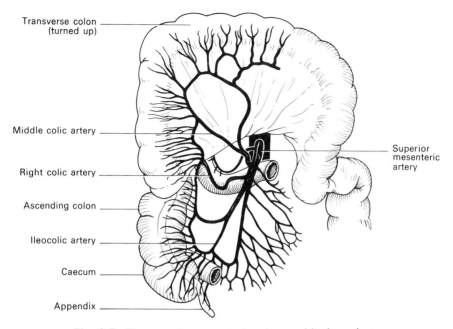

Transverse colon
(turned up)

Middle colic artery

Right colic artery

Ascending colon

Ileocolic artery

Caecum

Appendix

Superior
mesenteric
artery

Fig. 9.7. The superior mesenteric artery and its branches.

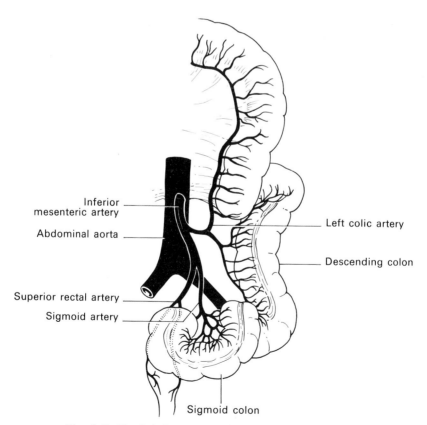

Inferior
mesenteric artery

Abdominal aorta

Superior rectal artery

Sigmoid artery

Left colic artery

Descending colon

Sigmoid colon

Fig. 9.8. The inferior mesenteric artery and its branches.

The iliac arteries

The aorta terminates in front of the body of the 4th lumbar vertebra by dividing into the *common iliac arteries*, right and left. Each common iliac artery runs downwards and outwards and divides into the internal iliac artery and the external iliac artery.

The *internal iliac artery* is the artery to the structures inside the pelvis and to the gluteal region. It gives off branches which supply:

the bladder,

the lower end of the rectum,

the uterus and vagina (in women),

the gluteal muscles.

The *external iliac artery* continues in the line of the common iliac artery to pass under the inguinal ligament and become the femoral artery.

The *femoral artery* is the principal artery to the leg (Fig. 9.9). It is a continuation of the external iliac artery, begins at the inguinal ligament, runs down the inner side of the thigh and then goes backwards on the medial side of the femur through an opening in the adductor magnus muscle to become the popliteal artery. It gives off branches to the muscles of the thigh and to the femur.

The *popliteal artery*, the continuation of the femoral artery, runs downwards in the popliteal space behind the knee to divide into the anterior tibial artery and the posterior tibial artery. The *anterior tibial artery* runs down the front of the leg to the dorsum of the foot. The *posterior tibial artery* runs down the back of the leg, then in the groove on the inner side of the ankle behind the medial malleolus to the sole of the foot.

The pulse

The *pulse* is a wave transmitted through the arteries as a response to the ejection of blood from the heart into the aorta. It is not due to the passage of blood along an artery, for this follows the wave. It is most easily felt when an artery is compressed lightly against a bone.

Places to feel pulse
radial artery at wrist
superficial temporal artery in front of ear
dorsalis pedis artery on dorsum of foot

Clinical features

The pulse varies in speed, regularity and force as it reflects changes in the heartbeat.

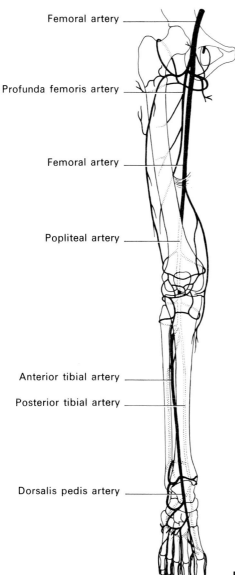

Femoral artery

Profunda femoris artery

Femoral artery

Popliteal artery

Anterior tibial artery

Posterior tibial artery

Dorsalis pedis artery

Fig. 9.9. The blood supply to the leg.

Pulse increased in speed in:
    physical exercise,
    anxiety,
    hyperthyroidism,
    anaemia.

Pulse decreased in speed in:
    rest,
    heart block.
Pulse irregular in:
    sinus arrhythmia,
    frequent extra-systoles,
    atrial fibrillation.
Pulse strong in:
    raised BP,
    excitement,
    hyperthyroidism.
Pulse weak in:
    shock
Pulse absent in:
    cardiac arrest,
    complete obstruction of an artery.

Blood pressure

Blood pressure (BP) is the pressure exerted by blood within a blood vessel. It is the result of:
(a) the cardiac output,
(b) the resistance to the flow of blood set up by blood vessels, mainly by the calibre of the arterioles.

It is highest in the aorta and diminishes the further the blood has to flow. It is raised by emotion and exercise and falls during sleep. It is measured in millimetres of mercury (mmHg). Two measurements are used:

    the *systolic pressure*: the pressure at cardiac systole,
    the *diastolic pressure*; the pressure at cardiac diastole.

Normal pressures (in mmHg)
    in aorta and large vessels: systole 120, diastole
    80
    in small arteries: systole 110, diastole 70
    in arterioles: systole 40

THE MICROCIRCULATION

The microcirculation is the circulation of the blood through blood vessels too small to be seen with the naked eye (Fig. 9.10).

> *Blood vessels of microcirculation*
>   arterioles
>   thoroughfare vessels
>   capillaries
>   sinusoids
>   venules

Arterioles

The arterioles are blood vessels with a relatively thick wall of smooth muscle.

The muscle of the arteriolar wall can be contracted or relaxed. Normally it is in a state of partial contraction. Contraction causes a constriction of the calibre

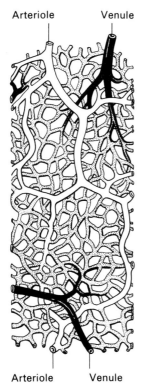

Arteriole   Venule

Arteriole   Venule

Fig. 9.10. A network of capillaries with the blood entering through the arterioles and leaving through the venules.

of the vessel. If the constriction is a local one, the supply of blood to the tissue or organ is reduced. If there is a generalized constriction, the BP is raised. The arterioles mainly involved are those in:

    the splanchnic area (i.e. of abdominal organs and tissues),
    the skin.

The degree of contraction of the arterioles is controlled by:

    nervous impulses,
    chemical substances in the blood.

NERVOUS CONTROL

Sympathetic nerves are the principal controlling nerves. Impulses passing along them arise in the vasomotor centre and cause contraction of the arteriolar muscle.

The *vasomotor centre* is a nucleus of cells in the medulla oblongata of the brain. It is to some extent controlled by: (a) centres in the cerebral cortex, (b) a temperature regulating centre in the hypothalamus.

The mechanisms by which adjustment of BP are made are:
(a) An increase in arterial pressure stimulates certain receptor cells in the aorta and carotid bodies (at the bifurcation of the common carotid arteries); these receptor cells are sensitive to the stretching of the arterial wall produced by a rise in BP.
(b) Stimuli are sent from these cells to the vasomotor centre.
(c) The number of stimuli sent to the arteriolar muscle is reduced.
(d) In consequence the arteriolar muscle relaxes, the calibre of the arterioles increases and the BP falls.
(e) If the BP falls too low, the receptor cells cease sending messages to the vasomotor centre; the number of stimuli sent to the arterioles increases, and the BP rises again.

CHEMICAL CONTROL

Vasoconstriction can be produced by:
    adrenaline,
    noradrenaline,
    angiotensin.
Vasodilatation can be produced locally by:
    carbon dioxide,
    adenosine monophosphate (AMP),
    bradykinin.

Thoroughfare vessels and capillaries

Thoroughfare vessels are thin-walled vessels which run directly from arteriole to venule. The *capillaries* (Fig. 9.10) are a mesh of smaller vessels opening off these thoroughfare vessels. The entrances to the capillaries are encircled by sphincters of smooth muscle. When these sphincters are open the blood goes into the capillaries. When they are closed the blood goes straight from arteriole to venule, missing the capillaries.

The capillaries are thin-walled tubes, about 6–8 μ in diameter, formed of a continuous layer of endothelial cells lying on a basement membrane. They run through the tissue spaces between cells.

Structures in which there are no capillaries
cornea of the eye
lens of the eye
epidermis of the skin

The BP in the capillaries falls from about 30 mmHg at the arterial end to about 10 mmHg at the venous end. Capillary pressure becomes raised when the arterioles dilate and the pre-capillary sphincters relax and more blood goes into the capillaries.

Capillaries open and close at a rate of 6–12 times a minute as a result of an opening and closing of the pre-capillary sphincters. The sphincters relax in response to any increase in the amount of carbon dioxide and lactic acid in the blood or any decrease in the amount of oxygen. This relaxation ensures that more blood gets to the tissues when there is an increase in metabolic activity. The sphincters to the capillaries of the skin relax in response to an increase in body temperature, and the increased circulation through the skin capillaries causes a fall in temperature.

The majority of capillaries in the body are closed at any one time. If all the capillaries were open at the same time, the body would bleed to death into its own capillaries. The closed capillaries of any organ form a reserve which can be called upon should an increase in blood flow become necessary.

The function of the capillaries is the exchange of oxygen and carbon dioxide, of nutrients, of fluid, and of waste products between the blood and the tissue-fluid surrounding cells. The velocity of blood flow through the capillaries is only about 0.5 mm/second, and this provides a relatively long time for these exchanges to take place. The capillary wall acts as a fairly permeable membrane. Water passes freely through the membrane, and so do molecules with a molecular weight below 5000. Molecules with a greater molecular weight may pass through and it has been suggested that there are pores between cells in the capillary wall through which they pass. White blood cells appear to be able to wriggle through the wall between cells.

Sinusoids

Sinusoids are present in the spleen, liver, bone marrow and endocrine glands. They are three to four times larger than capillaries and are partly lined with cells of the reticulo-endothelial system. Where there are sinusoids, the blood is in more direct contact with cells, and exchanges do not have to take place through tissue-spaces.

Veins and venules

The *venules* are the small veins formed by the union of capillaries.

Veins are formed by the union of venules. They have three walls imperfectly demarcated one from another:

an inner smooth layer of endothelial cells,
a middle layer of muscle and elastic fibres,
an outer layer mostly of collagen fibres.

Valves are present in many veins (Fig. 9.11). They are formed of two or three flaps, give the vein a beaded appearance, and prevent blood going in the wrong

Fig. 9.11. A valve within a vein, open (right) and closed (left).

direction. Some of the valves in the more central veins are imperfectly formed and not very competent.

Veins are supplied with sympathetic nerve fibres, not with any parasympathetic nerves.

Clinical feature
Varicose veins are dilated veins with incompetent valves. They commonly occur in the lower limbs, but can occur elsewhere, e.g. haemorrhoids at the anus.

The veins have two functions: to transport blood and to act as a reservoir.

Transport function
The veins are the tubes through which blood is returned to the heart.

The velocity of the blood in them is greater than in the capillaries, but not as great as that in the arteries. The resistance to the flow of blood in them is small, and the BP falls from about 10 mmHg in the venules to about zero in the great veins near the heart.

The movement of the blood along the veins is achieved by:
(a) gravity in veins above the heart,
(b) a muscle pump working in the muscles of the leg and the abdomen. Contractions of muscles compress the veins in the fascial planes between them; blood is forced upwards and cannot get back because of the valves. Similarly in the abdomen, contractions of the viscera may pump blood along their veins.

Reservoir function
The veins provide the largest available space for blood, a space in which it can be stored at a low pressure. Much of the blood—possibly more than 70 per cent—is in the veins at any one time. Veins can stretch to take in more blood without the pressure in them being raised. The reservoir of blood becomes less, by contraction of the vein wall, when more blood is required in the active tissues, such as the muscles during physical exercise.

VENOUS DRAINAGE

Head and neck
The veins of the head and neck (Fig. 9.12) mostly enter the *internal jugular vein*.

The *internal jugular vein* starts at the inferior surface of the skull and runs deep in the neck, under cover of the sternomastoid muscle and parallel with the internal and common carotid arteries. It terminates by joining the subclavian

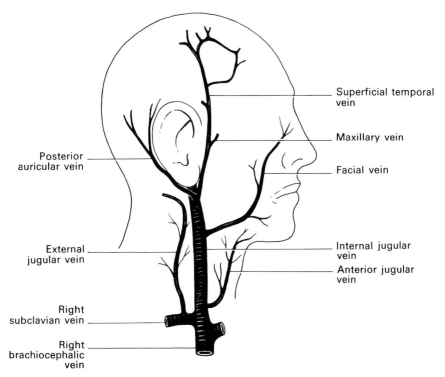

Fig. 9.12. The venous drainage of the face and neck.

vein from the arm to form the *brachiocephalic vein*. The left brachiocephalic vein crosses from left to right, just above the arch of the aorta and in front of the big vessels that arise from it, to join the right brachiocephalic vein. The two brachiocephalic veins unite on the right side of the neck to form the superior vena cava.

Brain
The veins do not correspond with the arteries to the brain, but form *venous sinuses*, which are channels composed of dura mater. The important sinuses are:

the *superior sagittal sinus*, which runs from front to back, just under the skull, in the midline,

the *inferior sagittal sinus*, which runs in the posterior part of the lower free border of the falx cerebri,

the *transverse sinus*, which runs laterally and as the *sigmoid sinus* passes through a hole in the base of the skull to be continued as the internal jugular vein,

the *cavernous sinus*, on either side of the body of the sphenoid bone; the internal cartoid artery and some cranial nerves run through this sinus.

Arm
Veins on the dorsum of the hand communicate with veins in the palms. Large

veins run up the front and back of the forearm, and with the arteries. There is a large *cephalic vein* on the outer side and a *basilic vein* on the inner side of the forearm; a communicating vein between them in front of the elbow is commonly used for venepuncture. The *axillary vein* is the continuation in the axilla of the basilic vein and in the neck becomes the subclavian vein. Blood from the arm passes through the subclavian vein into the brachiocephalic vein and so into the superior vena cava.

Superior vena cava
The superior vena cava is formed by the union of the two brachiocephalic veins on the right side of the neck. It receives blood from the head, neck, arms and upper part of the thorax. It runs downwards in the thorax to enter the right atrium of the heart.

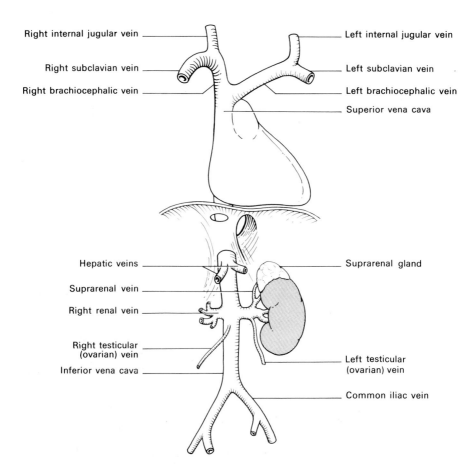

Right internal jugular vein — Left internal jugular vein
Right subclavian vein — Left subclavian vein
Right brachiocephalic vein — Left brachiocephalic vein
— Superior vena cava
Hepatic veins — Suprarenal gland
Suprarenal vein
Right renal vein
Right testicular (ovarian) vein — Left testicular (ovarian) vein
Inferior vena cava — Common iliac vein

Fig. 9.13. The superior and inferior venae cavae and their tributaries.

Leg

Veins run in the legs both with the arteries and independently. The *long saphenous vein* begins on the dorsum of the foot, passes in front of the internal malleolus and up the inner side of the leg and thigh. At the upper end of the thigh, it goes through a hole in the deep fascia to join the femoral vein. The *short saphenous vein* at the back of the calf passes into the popliteal fossa behind the knee where it joins the *popliteal vein*, formed from deep veins which have come up the leg with the arteries. The popliteal vein passes upward and to the front of the thigh to become the femoral vein. The *femoral vein* is on the inner side of the femoral artery in the femoral canal, in the groin, just below the inguinal ligament; it passes under the inguinal ligament to become the external iliac vein.

The *external iliac vein* is the continuation of the femoral vein. The *internal iliac vein* drains structures in the pelvis. The *common iliac vein* is formed by the union of the internal and external iliac veins.

Inferior vena cava

The inferior vena cava (Fig. 9.13) is formed by the union of the two common iliac veins. Below, it runs on the right of the aorta. It is a big vein and receives the right and left renal veins, the lumbar veins and some others. At the upper part of the abdomen, it passes to the right, away from the aorta, passes behind the liver and there receives the two hepatic veins from the liver, passes through an opening in the diaphragm and opens into the right atrium of the heart.

Special circulations

> *Special circulations*
> pulmonary circulation
> coronary circulation
> cerebral circulation
> portal system

PULMONARY CIRCULATION

The pulmonary circulation of blood passes from the right ventricle, through the pulmonary arteries and the capillaries which surround the alveoli in the lungs, and back to the left atrium of the heart.

The pulmonary artery arises at the upper end of the right ventricle. The opening between ventricle and artery is guarded by the pulmonary valve. The artery is short and divides into right and left branches which enter the respective lungs at their roots.

The pulmonary artery is the only artery to contain deoxygenated blood. The *pressure* within the artery is dependent upon:

(a) the output of the right ventricle,
(b) the resistance to the flow of blood through the pulmonary vessels.

The systolic pressure is 15–20 mm of mercury. The energy required to circulate the blood through the pulmonary circulation is much less than that required to circulate it through the systemic circulation; and the right ventricle has a muscular wall only about one-third as thick as that of the left ventricle.

CORONARY CIRCULATION

The coronary circulation flows through the coronary arteries, cardiac capillaries, and back to the heart through some small veins which open into the right atrium. The heart has a great need for oxygen and removes, even at rest, a large amount of the oxygen in the blood passing through the cardiac capillaries. In exercise there is a greater demand for oxygen. To meet it:
(a) the arterial blood pressure increases,
(b) the coronary arteries dilate.

CEREBRAL CIRCULATION

The circulation through the brain is kept fairly constant whatever changes there may be in the blood pressure. This is done by means of an auto-regulating mechanism. The autonomic nervous system has little control over the cerebral arteries, which have little muscle in their walls. It is important that the brain gets all the blood it needs; consciousness is lost if the blood supply to the brain is stopped for 5 seconds, and if it is not quickly and fully resumed brain cells start to die.

THE PORTAL SYSTEM

The portal system is composed of veins which drain the alimentary tract within the abdomen, the spleen and the pancreas. It is composed of the portal vein and its tributaries.

The tributaries are:

the *splenic vein*, which receives blood from the spleen and stomach,

the *inferior mesenteric vein*, which receives blood from the descending colon and upper part of the rectum, and opens into the splenic vein,

the *superior mesenteric vein*, which receives blood from the rest of the large intestine, the small intestine and the stomach.

The *portal vein* is formed by the union of the splenic vein with the superior mesenteric vein. It runs upwards and to the right towards the liver and divides into right and left branches, which enter the liver. These branches divide into smaller branches, and eventually the blood enters sinusoids around the liver cells. The blood then passes into larger vessels which eventually become the two hepatic veins. The *hepatic veins* (Fig. 9.14) arise from the back of the liver and enter the interior vena cava as it passes at the back of the liver.

The portal system ensures that all the chemical products of digestion (amino acids, lipids, carbohydrates, vitamins, minerals) are transported directly to the liver without having to circulate first through the general circulation.

Sinusoids   IVC   Hepatic veins

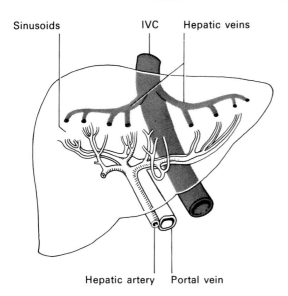

Fig. 9.14. The blood
supply and drainage of
Hepatic artery   Portal vein   the liver.

Anastomoses between the general circulation and the portal circulation occur
at:
  the cardiac end of the stomach,
  the lower end of the rectum,
  the umbilicus.
The blood vessels at these places can become dilated if there is an obstruction
to blood flow through the liver.

10
The Respiratory System

> *Respiratory system*
> nose
> pharynx
> larynx
> trachea
> bronchus, right and left
> lung, right and left
> pleura, right and left

Nose

The nose (Figs 10.1, 10.2) is composed of the external nose and the nasal cavities behind the external nose.

The *external nose* is composed of cartilage below and the nasal bones above, covered on the outside with skin and on the inside with mucous membrane.

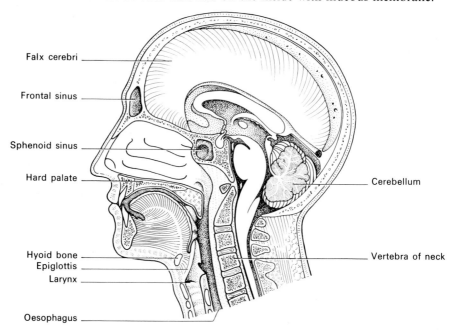

Falx cerebri

Frontal sinus

Sphenoid sinus

Hard palate

Cerebellum

Hyoid bone
Epiglottis
Larynx

Vertebra of neck

Oesophagus

Fig. 10.1. Section through the head.

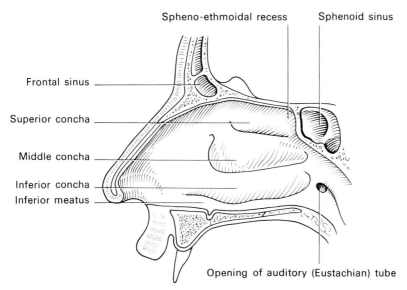

Fig. 10.2. The inside of the nose.

The *nasal cavities* extend from the nostrils in front to the posterior apertures of the nose, which open into the nasopharynx behind. They are lined with mucous membrane.

The *nasal septum* separates the two nasal cavities. It is a thin structure composed of bone and cartilage, is often bent to one side or the other, and is lined on both sides with mucous membrane.

The lateral wall of the nasal cavity is formed by parts of the maxilla, palatine and sphenoid bones.

The *upper, middle and lower conchae* (turbinate bones) are three delicate curved bones attached to the lateral wall and projecting into the nasal cavity. They are covered with mucous membrane.

The floor of the nasal cavity is formed by the maxilla and palatine bones.

The roof of the nasal cavity is a narrow chink formed by the frontal and sphenoid bones. The *olfactory mucous membrane*, in the roof and adjacent parts of the nasal cavity, contains special nerve cells in which smells are detected; from these cells nerve fibres pass through the cribriform plate of the frontal bone and into the olfactory bulb of the 1st cranial (olfactory) nerve.

Paranasal sinuses
  sphenoid
  ethmoid
  frontal
  maxillary

The *paranasal sinuses* (Fig. 10.3) are those spaces in the cranial bones which communicate through openings into the nasal cavity. They are lined with mucous membrane continuous with that of the nasal cavity.

Openings into the nasal cavities
 nostrils
 sphenoid sinus, above the superior concha,
 ethmoid sinus, by several openings between the superior and middle conchae and between the middle and inferior conchae,
 frontal sinus, between the middle and inferior conchae,
 maxillary antrum, between the middle and inferior conchae,
 nasolacrimal duct, below the inferior concha.
 At the back, the nasal cavity opens into the nasopharynx through the posterior nasal apertures.

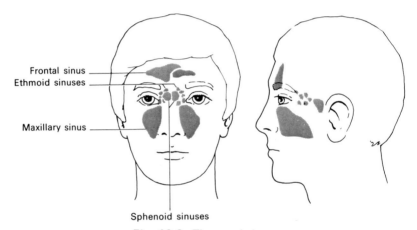

Frontal sinus
Ethmoid sinuses
Maxillary sinus
Sphenoid sinuses

Fig. 10.3. The nasal sinuses.

Nasopharynx

The nasopharynx lies immediately behind the nasal cavities, beneath the base of the skull and in front of the 1st and 2nd cervical vertebrae. It opens in front into the nasal cavities and below into the oral pharynx. The auditory (Eustachian) tubes open into its lateral wall on each side. The *nasopharyngeal tonsil* is a pad of lymph tissue in the posteriosuperior wall of the nasopharynx.

Oropharynx

The oral pharynx is common to the respiratory and alimentary systems as food enters it from the mouth and air from the nasopharynx and lungs.
 It is continuous below with the laryngeal pharynx, that part of the pharynx that lies immediately behind the larynx, and with the upper end of the oesophagus.

Inspired air is warmed, moistened and filtered as it passes through the nasal cavities.

Clinical features

The nasopharynx can be blocked by a common cold. The nasal septum is commonly deviated to one or other side. Enlargement of the conchae is common and can block a nasal cavity.

Inflammatory processes can spread from the nasal cavity into any of the nasal sinuses. *Sinusitis* is inflammation of a sinus. Drainage of an infected maxillary sinus into the nose is hindered by the opening of the sinus being high up in the wall of the nasal cavity.

Adenoids are an enlarged nasopharyngeal tonsil. By blocking the posterior nasal apertures they can cause mouth-breathing. By blocking the adjacent openings of the auditory (Eustachian) tubes they can prevent air getting up the tube into the middle ear.

Larynx

The larynx (Figs 10.4, 10.5, 10.6) is a complete structure of:
    cartilages: thyroid cartilage,
             epiglottis,
             cricoid cartilage,
             2 arytenoid cartilages.

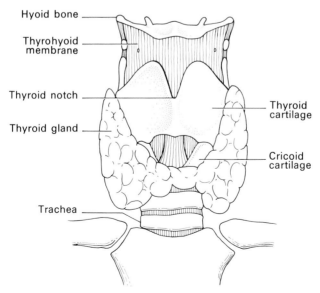

Fig. 10.4. Structures in the front of the neck.

membranes: connecting the cartilages with one another and with the hyoid bone.

mucous membrane.

vocal cords.

muscles acting on the vocal cords.

The larynx is situated in the midline of the front of the neck, deep to the skin, thyroid gland and some small muscles, and in frnt of the laryngeal pharynx and upper part of the oesophagus.

CARTILAGES

Thyroid cartilage (Fig. 10.5)

A V-shaped piece of cartilage, with the V projecting forwards in the neck as Adam's apple. The posterior border ends above in a superior horn, a projection

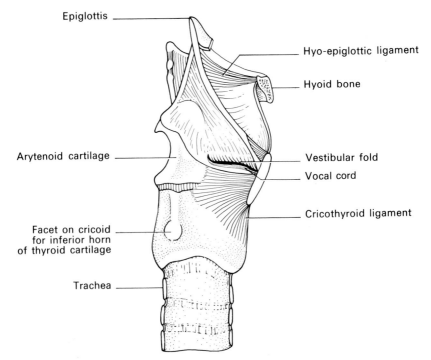

Fig. 10.5. The larynx seen from within. The vocal whose vibrations make sound, extend from the thyroid cartilage in front to the arytenoid cartilages behind.

to which the thyrohyoid ligament is attached, and below in a smaller horn which articulates with the outside of the cricoid cartilage.

The *thyrohyoid membrane* connects the upper border and superior horns to the hyoid bone.

The *cricothyroid membrane* connects the lower border with the cricoid cartilage.

Epiglottis
A leaf-shaped piece of cartilage which projects upwards behind the base of the tongue. It is attached to the back of the hyoid bone and the back of the V of the thyroid cartilage.

The *aryepiglottic folds*, running backwards from the sides of the epiglottis to the arytenoid cartilages, form the margins of the entrance to the larynx.

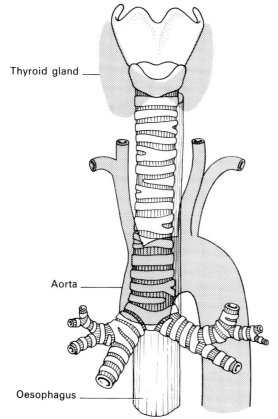

Thyroid gland

Aorta

Oesophagus

Fig. 10.6. The larynx, trachea and bronchi.

Cricoid cartilage
A signet-ring shaped piece of cartilage with the big part behind. It lies below the thyroid cartilage, being connected to it by the cricothyroid membrane. The inferior horns of the thyroid cartilage articulate with it on either side. The *cricotracheal membrane* connects its lower border with the 1st ring of the trachea.

Arytenoid cartilages
Two tiny pyramidal-shaped pieces of cartilage which sit on the base of the cricoid cartilage. The vocal cord on each side is attached posteriourly to the angle of the pyramid that projects forwards.

MUCOUS MEMBRANE

The larynx is lined mostly with respiratory epithelium, consisting of ciliated columnar cells.
The vocal cords are covered with squamous epithelium.

VOCAL CORDS

The vocal cords are two shelves of thin mucous membrane lying over the vocal ligaments, two fibrous bands stretched between the inside of the thyroid cartilage in front and the arytenoid cartilages behind.
The false vocal cords are two folds of mucous membrane just above the true vocal cords. They are not involved in voice production.

MUSCLES

Small muscles are attached to the arytenoid, cricoid and thyroid cartilages, which by contracting and relaxing can approximate and separate the vocal cords. The muscles are supplied by the X cranial (vagus) nerve.

Respiration
During quiet respiration the vocal cords are held slightly apart so that air can go up and down. During forced respiration the vocal cords are widely separated.

Phonation
Sounds are produced by the vibrations of the vocal cords during expiration. The sounds thus produced are modified by movements of the soft palate, cheeks, tongue and lips, and given resonance by the cranial air sinuses.

Clinical features
The larynx can become blocked by:
(a) a foreign body, e.g. a lump of food, a small toy,
(b) swelling of the mucous membrane, e.g. after inhaling steam or as an allergic reaction,
(c) infections, e.g. diphtheria,
(d) a tumour, e.g. a cancer of a vocal cord.

Trachea and bronchi

THE TRACHEA

The trachea (Fig. 10.6, p. 121) is a flexible tube about 10 cm long and 2.5 cm wide. It runs from the cricoid cartilage down the front of the neck and behind the manubrium of the sternum, to end at the level of the sternal angle (the junction of the manubrium with the body of the sternum) where it ends by dividing into a right and left bronchus. In the neck it is crossed in front by the isthmus of the thyroid gland and several veins. It is composed of 16-20 C-shaped pieces of cartilage connected by fibrous tissue. Its construction is such that it remains open whatever the position of the head and neck.

THE BRONCHI

The right and left bronchi (Fig. 10.6) run downwards and outwards from the bifurcation of the trachea to the root of their respective lungs. The right bronchus is wider, shorter and more vertical than the left. At the root of the lung the bronchus divides into smaller bronchi which enter the lung. Each bronchus is about half the calibre of the trachea and is composed similarly, on a smaller scale, of cartilages connected by fibrous tissue.

Clinical features
Tracheitis is inflammation of the trachea. *Bronchitis* is inflammation of the bronchi. *Foreign bodies* (e.g. small toys, fruit stones, pips) are more likely to enter the right than the left bronchus because of its size and direction.

    Tracheostomy is the making of a hole into the trachea. It may be neccessary:
(a) in the administration of an anaesthetic for certain operations,
(b) upper respiratory tract obstruction,
(c) for patients in coma,
(d) in some cases of paralysis,
(e) severe injuries of the jaw or face.

Lungs

Each lung (Fig. 10.7) is conical and has:
(a) an apex, which extends into the neck for about 2.5 cm above the clavicle,
(b) a costo-vertebral surface, moulded to the inside of the chest wall,
(c) a mediastinal surface, moulded to the pericardium and heart,
(d) a base, which lies on the diaphragm.

Root of lung
Root of lung is on its mediastinal surface.
The following enter or leave lung at it:
    bronchus and its main branches
    pulmonary artery
    pulmonary veins
    lymph vessels
    nerves

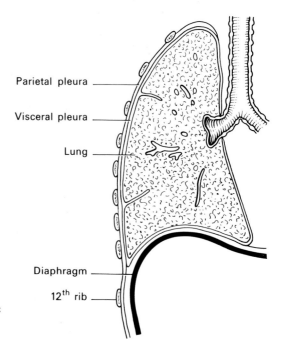

Parietal pleura _____

Visceral pleura _____

Lung _____

Diaphragm _____

12th rib _____

Fig. 10.7. Diagrammatic section through the lung and pleura.

The *right lung* (Fig. 10.8) is divided by two fissures into three lobes: upper, middle, lower.

The *left lung* is divided by one fissure into two lobes: upper, lower.

The *bronchus* of each side divides into main branches, one for each lobe of the lung. A *segment* of a lung is that area supplied by a main branch of the bronchus; each segment is a self-contained unit with its own blood supply. The right lung has ten segments, the left nine. Each segment is wedge-shaped with the thin edge of the wedge at the root of the lung.

Within its segment the main bronchial branch divides into smaller branches. A *bronchiole* (Fig. 10.9) is one of the smaller branches and one which has no cartilage in its wall. Each bronchiole divides into smaller branches. An *alveolar duct* is the smallest of these branches; each ends in a cluster of alveoli. An *alveolus* (Fig. 10.9) is a thin-walled air-containing sac, through whose wall the interchange of gases takes place. Each lung contains about 300 million alveoli. Tiny holes in the alveolar walls enable air to pass from one alveolus to another.

A *primary lobule* or *unit of lung* is a bronchiole with its clusters of alveoli.

BLOOD SUPPLY

Each *pulmonary artery*, bringing deoxygenated blood from the right ventricle of the heart, divides with each bronchus into branches for the lobes, segments and lobules. The terminal branches end in a network of capillaries on the surface of each alveolus.

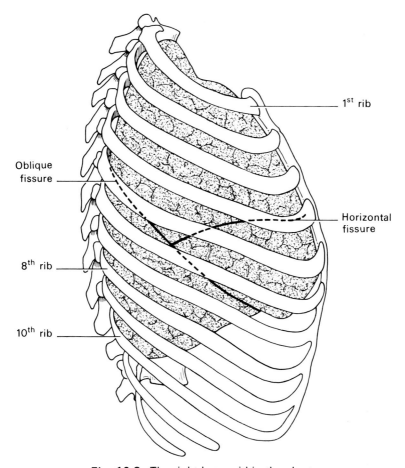

Fig. 10.8. The right lung within the chest.

1st rib

Oblique fissure

Horizontal fissure

8th rib

10th rib

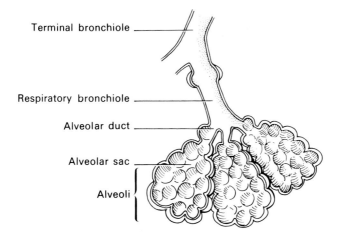

Fig. 10.9. The bronchioles and alveoli.

Terminal bronchiole

Respiratory bronchiole

Alveolar duct

Alveolar sac

Alveoli

This capillary network drains into progressively larger veins, which eventually form the *pulmonary veins*, two on each side, through which oxygenated blood passes to the left atrium of the heart.

Smaller *bronchial arteries* from the aorta supply the lung tissue with oxygenated blood.

LYMPH DRAINAGE

To lymph nodes at the bifurcation of the trachea, thence to lymph nodes around the trachea and in the mediastinum.

NERVE SUPPLY

(a) parasympathetic—via the vagus nerves; (b) sympathetic—from the sympathetic trunk.

Pleura

The *pleura* (Fig. 10.7, p. 124) is a thin, transparent membrane which encloses the lung in a double layer: a visceral layer, firmly adherent to the surface of the lung, and a parietal layer lining the inner surface of the chest wall. The two layers are continuous at the root of the lung. The *pleural cavity* is the space between the two layers. Adjacent surfaces are moist and capable of moving one on the other.

Clinical features
Pleurisy is inflammation of the pleura. A *pleural effusion* is fluid in the pleural cavity. *Empyema* is pus in the pleural cavity. *Pneumothorax* is air in the pleural cavity.

Mediastinum

The *mediastinum* is the area within the chest between the lungs. It is divided into superior and inferior mediastinum by an imaginary line drawn backwards from the sternal angle (the junction of the manubrium with the body of the sternum) to the 4th thoracic vertebra.

  Superior mediastinum contains:
  aortic arch and its branches
  superior vena cava and braciocephalic veins
  trachea   oesophagus
  thoracic duct   thymus gland or its remains
  vagus and phrenic nerves
  Inferior mediastinum contains:
anterior (in front of the heart)
  thymus or its remains
  connective tissue
middle
  heart and pericardium
  great blood vessels
posterior (behind the heart)
  descending aorta   thoracic duct
  oesophagus   vagus nerve

Respiration

> *Respiratory activities*
>   *ventilation*: the movement of air in and out of the
>   lungs
>   *diffusion*: the movement of oxygen and carbon
>   dioxide between the air in the alveoli and the
>   blood in the capillaries around them.
>   *transport*: the carriage of oxygen and carbon
>   dioxide by the blood
>   *tissue metabolism*: the exchange of oxygen and
>   carbon dioxide between blood and tissues

VENTILATION

Respiratory movements are inspiration and expiration.

In *inspiration* the diaphragmatic muscle contracts and the dome of the diaphragm descends; at the same time the external intercostal muscles contract and pull the chest wall slightly outwards. By these actions the space within the chest is enlarged, the pressure in the alveoli is decreased, and air enters the lungs.

In *expiration* the diaphragm and external intercostal muscles relax. The diaphragm rises, the chest wall falls in, and the space within the chest is diminished.

Normal quiet respiration occurs about sixteen times a minute. Expiration is followed by a slight pause. The depth and number of respiratory movements are mainly biochemically controlled, but this control can be modified by voluntary action in speaking, singing and whistling, and one can hold one's breath for about a minute.

In deep and forced respiration other muscles are brought into action—mainly the sternomastoid, scalenus anterior, pectoralis major and serratus anterior.

Lung volumes

Not all the air that is inspired reaches the lungs and not all the air in the alveoli is expelled in expiration.

The *tidal volume* is the volume of air inspired and expired in quiet breathing. At rest this amounts to about 400 ml. Of this amount:

(a) about 150 ml fill the *dead space* in the nasal passages, trachea, bronchi and bronchioles, and do not enter the alveoli,

(b) about 250 ml enter the alveoli, where they mix with about 3000 ml that have remained in them after a quiet respiration. About half of these 3000 ml can be expelled by voluntary effort after a normal quiet respiration has finished.

The *residual volume* (about 1500 ml) is the amount of air that cannot be expelled by forced respiration and remains in the alveoli.

The *vital capacity* is the amount of air (about 4500 ml) that can be expelled by voluntary effort after a deep inspiration.

All parts of the lungs are not equally ventilated. In the standing position, ventilation is greater at the bases than the apices.

DIFFUSION OF GASES

Gases pass almost instantaneously between the alveoli and the blood by diffusion. In this diffusion gas molecules pass from a place of high partial pressure to one of lower partial pressure.

The oxygen in the alveoli is at a higher partial pressure than that in the blood and so passes from the alveoli into the blood. The carbon dioxide in the blood is at a higher partial pressure than that in the alveoli and so passes from the blood into the alveoli.

The volume of gases transferred depends upon (a) the surface area of the alveoli, and (b) the thickness of the alveolar wall.

Clinical features
In some diseases of the lungs (e.g. pulmonary fibrosis, asbestosis) the alveolar wall becomes thickened. This impairs the diffusion of oxygen through the wall (but not of carbon dioxide).

Carbon dioxide is administered, in a mixture with oxygen, in order to stimulate respiration, e.g. during anaesthesia, in some cases of poisoning.

TRANSPORT OF GASES IN THE BLOOD

Oxygen is transported in the blood:
in *red cells*: oxygen combines with haemoglobin to form oxyhaemoglobin (oxyHb), which is bright red,
in *plasma*: some oxygen is carried dissolved in plasma.
Carbon dioxide is transported in the blood:
as bicarbonate:
(a) as sodium bicarbonate in the plasma,
(b) as potassium bicarbonate in red cells;
in solution,
combined with haemoglobin and plasma proteins.

Clinical features
Cyanosis is the purple colour of the skin and mucous membranes which is visible when the amount of non-oxygenated haemoglobin in the blood of the capillaries is more than 5 g per 100 ml of blood.
Common causes are:
(a) congenital deformities of the heart which cause the blood not to pass through the lungs,
(b) any disease of the lungs which prevents adequate diffusion of oxygen from the alveoli into the capillaries,
(c) a reduction of blood flow to a part of the body.

Carbon monoxide is more readily absorbed than oxygen by haemoglobin, the affinity of carbon monoxide for haemoglobin being 200–300 times greater than the affinity of oxygen. If haemoglobin takes up carbon monoxide, it cannot take up oxygen. This is the cause of death by carbon monoxide poisoning.

EXCHANGE OF GASES IN THE TISSUES

Oxygen
When oxygenated blood reaches the tissues, oxygen passes from the blood into the tissue-fluid because the partial pressure of oxygen in the blood is much higher than that in the tissue-fluid. From the tissue-fluid the oxygen passes into the cells according to their individual need for it.

Carbon dioxide
The carbon dioxide produced in the cells passes into the tissue-fluid. The partial pressure of carbon dioxide in the tissue-fluid is higher than that in the blood, and therefore carbon dioxide passes from tissue-fluid into the blood.

THE NERVOUS AND CHEMICAL CONTROL OF RESPIRATION

Respiration is a reflex act modifiable by voluntary control in the cerebral cortex.

Respiratory centres which stimulate inspiration and expiration are situated in the medulla oblongata of the brain.

The *carotid bodies* (at the bifurcation of each common carotid artery) and the *aortic bodies* (on the arch of the aorta) are tiny organs composed of nerve cells and blood vessels and connected by nerves to the respiratory centres in the medulla oblongata.

The carotid and aortic bodies and some centres in the medulla oblongata are sensitive to changes in the carbon dioxide tension and hydrogen ion concentration in the blood passing through them. An increase in CO_2 and a fall in pH causes the respiratory centre to send impulses to the muscles of respiration in order to stimulate them to contract more and more often so that CO_2 is expelled from the lungs and the pH restored to normal.

Efferent impulses are transmitted down the phrenic and intercostal nerves to the diaphragm and intercostal muscles. At the same time impulses sent down the IX cranial (glossopharyngeal) nerve and the X cranial (vagus) nerve control and modify the action of the other efferent impulses.

Stimulation of sense organs in the nose and tracheobronchial tree produce cough reflexes and sneeze reflexes which modify ventilation in the lungs.

Clinical features
The lungs are subject to many diseases, e.g.
  infections: pneumonia, tuberculosis, viral infections etc.
  new growths: cancer of a bronchus rapidly invades lung tissue.

allergies: especially asthma, in which there is a spasm of the muscle of the bronchi and an accumulation of secretion in the bronchi.

dust diseases: due to the inhalation of harmful dust produced in some industrial processes, e.g. silica, asbestos.

physical damage to lung tissue, e.g. emphysema, in which the alveoli and smaller bronchioles are distended and lose their elasticity.

A *failure of respiratory function* can occur in any severe pulmonary disease. Inadequate oxygenation of the blood produces cyanosis and impaired mental functioning.

11
Blood

Blood is the transport medium of the body. The total amount of blood in a man of 70 kg is 5 litres.

Composition of blood
    red cells
    white cells
    platelets
    plasma

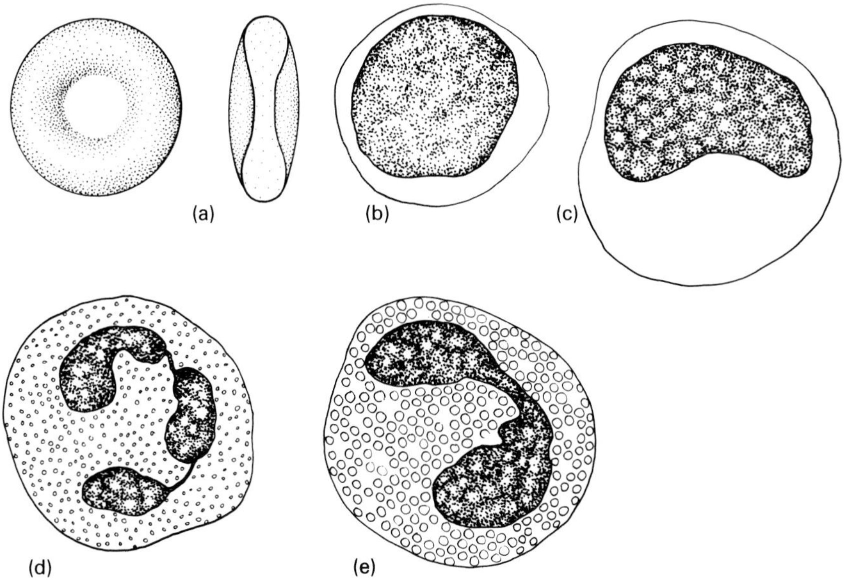

(a)   (b)   (c)

(d)   (e)

Fig. 11.1. Blood cells: (a) red cell in full face and on section, (b) lymphocyte, (c) monocyte, (d) granulocyte with three lobes, (e) granulocyte with two lobes.

RED CELLS (erythrocytes, red corpuscles)

Red cells are biconcave discs with a diameter of about 8.6 μm. The biconcavity allows for rapid movement of oxygen in and out of the cell as it provides the shortest distance between the membrane and the contents of the cell. They do not have a nucleus. They are composed of:

an outer membrane,

haemoglobin (Hb), an iron-containing protein,

carbonic anhydrase, an enzyme involved in the transport of carbon dioxide.

The normal range of numbers of red cells is:

males $4.5-6.5 \times 10^{12}/l$

females $3.9-5.8 \times 10^{12}/l$

Development

Before birth: in yolk sac; subsequently in the spleen, liver and red bone marrow. *After birth*: in red bone marrow, which with increasing age becomes limited to the sternum, vertebrae, and heads of femur and humerus.

Normoblasts are nucleated cells in the bone marrow from which red cells develop with the extrusion of the nucleus from the cell. *Reticulocytes* are primitive red cells in which a fine network can be seen with special staining methods; they normally form less than one per cent of circulating red cells. The bone marrow turns out 2 million red cells a second to replace those lost.

Substances necessary for red cell development	
iron	vitamin B_{12}
folic acid	certain amino acids
cobalt	copper

Life span

Red cells live for 74–154 days. At this age their enzyme system appears to fail, the cell membrane stops functioning adequately, and they are destroyed by cells of the reticulo-endothelial system. Haemoglobin is broken down into:

haem, which contains the iron; most of the iron is retained in the body and used in the manufacture of new red cells,

porphyrin, which is broken down into bilirubin; the bilirubin circulates in the plasma and is removed by cells of the liver as the blood circulates through that organ. It is excreted in the bile.

Transport of oxygen

Haemoglobin has an affinity for oxygen. 1 g of haemoglobin will combine with 1.34 ml of oxygen. Normally a person has 14.5 g of haemoglobin in each 100 ml of blood. Each 100 ml of blood is therefore carrying 20 ml of oxygen.

As blood passes through the capillaries of the lungs, oxygen is taken out of the air in the alveoli into the red cells. As the blood passes through the capillaries of tissues that require oxygen, oxyhaemoglobin gives up oxygen.

Oxyhaemoglobin is haemoglobin combined with oxygen and is bright red. Reduced haemoglobin is haemoglobin without oxygen and is dark blue, almost black.

Transport of carbon dioxide

Some carbon dioxide is transported in red cells in addition to that transported in the plasma. It too can combine with haemoglobin. The process of taking it up is accelerated by the enzyme carbonic anhydrase present in red cells.

Buffer

Haemoglobin is a part of the buffer system controlling the pH of the body.

Anaemia

Anaemia is present when there is insufficient haemoglobin to carry oxygen in adequate amounts to the cells of the body. The causes of anaemia are:

Loss of blood: by haemorrhage, menstruation.

Impairment of red cell formation:

(a) *aplastic anaemia*: with a failure of bone marrow activity, an inadequate number of red cells are produced,

(b) *deficiency anaemias*: the result of a shortage of one of the substances necessary for red cell formation. The shortage can be of any one of three substances: iron, vitamin B_{12}, folic acid,

(c) *increase in the rate of red cell destruction*: this occurs in malaria (due to malarial parasites invading red cells), in lead poisoning, in the presence of an abnormal haemoglobin, as the result of the action of some drugs, as the result of an inherited metabolic defect etc.

Blood groups

People are divided into four main blood groups—AB, A, B and O.

Proportions of people in different blood groups	
group AB	5 per cent
group A	40 per cent
group B	10 per cent
group O	45 per cent

Red cells have on their surface certain antigens. These antigens are genetically determined:

group AB cells have A and B antigens,
group A cells have A antigen,
group B cells have B antigen,
group O cells have no antigens.

A person's plasma contains antibodies against the antigens lacking on his cells.

A *group A person* has therefore:

(a) A antigens on his red cells,

(b) anti-B antibodies in his plasma.

A *group B person* has:

(a) B antigens on his red cells,

(b) anti-A antibodies in his plasma.

If the blood of a group A person were injected into the blood of a group B person, the anti-A antibodies of B would clump the donated cells of A and cause smaller blood vessels to become blocked.

Similarly:

a group AB person has none of these antibodies in his plasma,

a group O person has anti-A and anti-B antibodies in his plasma.

A person of group O cannot be a 'universal donor' because:

(a) anti-A and anti-B antibodies in his plasma are sometimes so potent that they can agglutinate the cells of a non-O recipient,

(b) group O can contain rare antigens which would cause trouble if injected into a non-O person.

For similar reasons a person of group AB cannot be a 'universal recipient'.

RHESUS BLOOD GROUPS

Rhesus blood groups are so-called because blood from rhesus monkeys was used in the original experiments. Some people were found to possess a particular antigen and some not to possess it.

Rhesus groups	
Rh-positive	85 per cent of people
Rh-negative	15 per cent of people

The Rh-positive factor is inherited as a dominant and the Rh-negative factor as a recessive. If a person is Rh-positive, either one or both of his parents must be Rh-positive. If a person is Rh-negative, both his parents must be Rh-negative. There are several factors involved and of these D is the important one.

Clinical features

If a Rh-negative woman is made pregnant by a Rh-positive man, the child may be Rh-positive. When this happens there is a danger of rhesus factor D incompatibility developing between the Rh-negative mother and the Rh-positive fetus. This is the main cause of serious haemolytic disease of the newborn. The disease is not, however, likely to develop in a first incompatible pregnancy unless the mother has any time previously received a transfusion of Rh-positive blood.

During labour small leaks (1–5 ml) of fetal blood into the maternal circulation are common, and as little as 0.5 ml of Rh-positive blood can cause the mother to develop anti-D antibodies. In further incompatible pregnancies anti-D antibodies pass through the placenta to coat the red blood cells of the fetus. This can cause the fetus to die in utero of anaemia and cardiac failure. Affected live born babies are anaemic and can develop a severe jaundice.

Prevention
(a) Rh-negative women should never be transfused with Rh-positive blood.
(b) If there is evidence of a feto-maternal leak of blood after the 1st incompatible pregnancy, the mother is given anti-D globulin within 36 hours of delivery. This prevents the development of anti-D antibodies.

Treatment
(a) An affected fetus can be given an intra-uterine transfusion of red cells.
(b) An affected live born baby is given an exchange transfusion.

Rh-negative people, men and women, should not be transfused with Rh-positive blood, for antibodies can develop in them and a second transfusion can cause a haemolytic reaction.

WHITE CELLS (leucocytes)

Types of white cells
  granulocytes
  lymphocytes
  monocytes

At birth the total number of white cells is up to 25 000 per mm^3. The number falls during childhood.

Numbers of white cells in adult life

total $4.0–11.0 \times 10^9/l$

made up of:	$\times 10^9/l$	per cent
granulocytes:		
neutrophils	2.5–7.5	40–75
eosinophils	0.04–0.44	1–6
basophils	0–0.10	0–1
lymphocytes	1.5–3.5	20–45
monocytes	0.2–0.8	2–10

Granulocytes

Granulocytes have small granules in their protoplasm. The staining properties of these granules divide them into three groups:
   neutrophils: granules do not stain,
   eosinophils: granules stain red with acid dyes,
   basophils: granules stain blue with basic dyes.
   Granulocytes are about 10–12 μm in diameter, being thus larger than red cells. As it gets older a granulocyte's nucleus becomes divided into several lobes: hence the name polymorphonuclear leucocytes (usually abbreviated to polymorphs).
   They develop in bone marrow and are discharged into the circulation as required.

Lymphocytes

Lymphocytes have a large round or slightly indented nucleus, which occupies most of the cell. They develop in lymph tissue. They vary in size from 7 to 15 μm.

Monocytes

Monocytes are large cells, up to 20 μm in diameter, with an oval or kidney-shaped nucleus. They are formed in bone marrow.

Neutrophils can absorb and destroy micro-organisms. Lymphocytes are involved in immunity reactions (see Chapter 24). The functions of the other white cells are not definitely known.

Clinical features
The number of white cells can be increased or decreased.
Leucocytosis: an increase in the number of white cells. It can occur in:
   many infections, e.g. pneumonia,
   leukaemia,
   rapidly advancing malignant disease,
   after haemorrhage or severe injury.
Lymphocytosis: an increase in the number of lymphocytes. It can occur in:
   some infections, e.g. whooping cough,
   lymphatic leukaemia.
Eosinophilia: an increase in the number of eosinophil granulocytes. This occurs in some parasitic infections, e.g. toxocariasis.
Leukaemia is a disease in which there is an over-production of white cells. There are several types of leukaemia, classified according to (a) whether it is acute or chronic, (b) the type of cell increased:
   lymphatic leukaemia: an increase in lymphocytes,
   monocytic leukaemia: an increase in monocytes,

myeloid leukaemia: the presence of myeloid cells, the primitive cells from which granulocytes are formed.

Leucopenia: a reduction in the number of leucocytes. It is usually a neutropenia, i.e. a reduction in the number of neutrophil granulocytes. It occurs in:

some infections, e.g. influenza,

aplastic anaemia: bone marrow function is impaired or stopped,

some drug reactions, e.g. amidopyrine, thiouracil group.

Agranulocytosis: a total of near-total lack of granulocytes, a condition associated with severe symptoms. The causes are the same as those of leucopenia.

PLATELETS

Platelets (thrombocytes) number between 150 and 400×10^9/litre (150 000–400 000/μl). They are round or oval, biconvex, non-nucleated discs. They are portions of some large cells of the bone marrow, and live for about 10 days. About 30–40 per cent of them are concentrated in the spleen; the rest circulate in the blood, keeping close to the endothelium (the innermost lining of a blood vessel).

Functions
to maintain the integrity of the endothelium
(how they do this is uncertain)
to control bleeding (see p. 139)

Clinical features

Thrombocytopenia is an abnormal reduction in the number of platelets. Causes are: (a) diminished bone marrow function due to leukaemia, some drugs (salicylates, sulphonamides etc), invasion of bone marrow by malignant disease; (b) excessive destruction in some autoimmune diseases; (c) shock and some cases of septicaemia. Haemorrhages occur when the number of platelets falls below 10×10^9/litre. A *petechia* is a small cutaneous haemorrhage. An *ecchymosis* is a large haemorrhage. *Purpura* is any condition in which multiple haemorrhages occur in the skin or subcutaneous tissue.

PLASMA

Plasma is the fluid part of blood. It forms about 5 per cent of body weight. Plasma provides:

(a) the medium in which the formed elements of the blood (red cells, white cells, platelets) circulate,

(b) transport of inorganic and organic substances from one organ or tissue to another.

Composition of plasma
Water: 91–92 per cent
Plasma proteins:
    albumin: forms the greater part of the protein
        content of the plasma; is produced in the liver
    globulins, α, β and γ; produced in the liver,
        lymphocytes and reticulo-endothelial cells. The
        immunoglobulins are globulins formed as part
        of the immunity reactions of the body
    fibrinogen: produced in the liver
    prothrombin: a precursor of thrombin
Inorganic constituents: sodium, potassium,
        calcium, magnesium, iron, iodine etc
Organic constituents: urea, uric acid, creatinine,
        glucose, lipids, amino acids, enzymes, hormones

The *special functions of plasma proteins* are:
(a) they maintain the osmotic pressure of plasma necessary for the formation and absorption of tissue-fluid,
(b) by combining with both acids and alkalis they act as buffers in maintaining the normal pH of the body,
(c) fibrinogen and prothrombin are necessary for the clotting of blood,
(d) immunoglobulins are essential in the defence of the body against infection.

Clinical features
Plasma is obtained by centrifuging off the cells of the blood. It is given intravenously:
(a) to restore blood volume, e.g. after fluid loss from extensive burns,
(b) to provide substances missing from a patient's blood; e.g. to provide factors I, VIII, IX and XI for patients who have not got them.
    Variations of plasma-protein content occur in:
    renal disease: the plasma albumin falls when there is a big leak of albumin through the renal glomeruli,
    chronic liver disease and *starvation*: the plasma albumin falls as a result of lack of protein and a failure of the liver to produce plasma proteins,
    infection: the amount of globulin is usually increased as part of the defence mechanisms of the body,
    some disorders of protein production, congenital or acquired; a decrease in globulin production is likely to cause an increased susceptibility to infection.

FLUIDITY

The *fluidity* of blood is achieved by several factors:
(a) the normal blood flow,
(b) a smooth intact endothelium lining the inside of all blood vessels,

(c) the presence of certain physiological anticoagulants which prevent clotting,
(d) the destruction of any fibrin formed.

HAEMOSTASIS

Haemostasis is the stopping of bleeding. The following are involved:
blood vessel,
platelets,
coagulation of blood.

Blood vessel
A damaged blood vessel becomes constricted by a nervous reflex so that the amount of blood escaping is reduced.

Platelets
Platelets stick to the damaged edges of the blood vessel and to one another. In this way an amorphous mass is produced which blocks the hole partly or completely. Serotonin is produced in this mass and produces a further constriction of the blood vessel.

Coagulation of blood
Blood clots over any damaged area in the blood vessel, strengthening the obstruction formed by the platelets and further closing the hole.

The coagulation of blood is a series of complex biochemical reactions involving at least twelve different components of plasma, which are numbered I to XII. Fibrin is the end-product.
The substances involved include:
prothrombin,
thromboplastin (produced by damaged cells and platelets),
calcium,
vitamin K,
plasma clotting factors,
fibrinogen (a plasma protein).
By their interreactions:
(a) prothrombin is converted into thrombin,
(b) thrombin acts with fibrinogen to form fibrin,
(c) fibrin is converted into insoluble fibrin,
(d) insoluble fibrin forms a web with red cells entangled in it, and this becomes the clot.

Fibrin contracts in time and *serum*, a pale yellow fluid is extruded from the clot.

Clinical features
Bleeding diseases, in which excessive or prolonged bleeding occurs, are due to:
an *interference with the coagulation cycle*: as in haemophilia (where factor VIII is missing), Christmas disease (where factor IX is missing), vitamin

K deficiency, and when anticoagulant drugs are being used for the treatment of thrombosis (the intravascular clotting of blood which occurs in certain conditions),

thrombocytopenia: a reduction in the number of platelets below the normal number,

damage or defect of the wall of blood vessels: due to a congenital deformity, scurvy, infections, metabolic disorders etc.

SPLEEN

The spleen (Fig. 11.2) is a soft purple organ about the size of one's fist. It lies in the upper left corner of the abdomen, under cover of the ribs (Fig. 11.3).

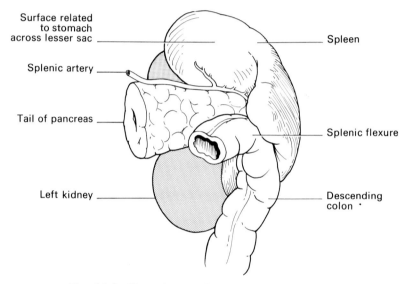

Fig. 11.2. The spleen and organs in contact with it.

It has a convex outer surface in contact with the diaphragm and a concave medial surface in contact with the stomach, splenic flexure of the colon and left kidney. The hilium is the place on the medial surface where blood vessels enter and leave the spleen. The tail of the pancreas reaches the hilium. The anterior border is notched. The spleen is enclosed within peritoneum.

Structure

The spleen is composed of:

a *capsule* of fibroelastic tissue,

lymph follicles: masses of lymph tissue, the same as that in lymph nodes,

red pulp: a network of connective tissue, red cells, white cells etc with many large sinusoids running through it.

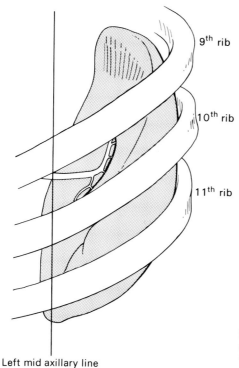

9th rib

10th rib

11th rib

Left mid axillary line

Fig. 11.3. The relation of the spleen to the 9th– 11th left ribs.

Blood supply
By the splenic artery, which arises from the coeliac artery.

Venous drainage
By the splenic vein, which is part of the portal system.

Functions of spleen
formation of red cells (in fetal life only)
destruction of old red cells
storage of iron from destroyed red cells
production of bilirubin from red cells
formation of lymphocytes in the lymph follicles
formation of immunoglobulin
removal of foreign particles from blood
reservoir (to small extent) of blood, which can be
   compressed out of it by contraction of capsule
   when required in circulation

Clinical features
Enlargement of the spleen occurs in some infections (especially malaria), leukaemia, lymphadenoma (Hodgkin's disease) etc. When it has enlarged about three

times, it becomes palpable below the left costal margin; the notch on its anterior border is a distinguishing palpable feature. As it enlarges further, it pushes downwards and to the right, to the umbilicus and right iliac fossa.

Splenectomy is surgical removal of the spleen. It is necessary for wounds of the spleen (which bleed profusely) and in some diseases.

RETICULO-ENDOTHELIAL SYSTEM

The reticulo-endothelial (RE) system consists of a number of cells of similar structure and with common functions situated in various organs and tissues.

RE cells present in:
spleen	liver
thymus	bone marrow
lymph nodes	blood vessel walls

The common function of all RE cells is the removal of particles of foreign matter, the destruction of worn-out red cells, and the destruction of some other cells.

Clinical feature
Cells of the RE system become filled with substances, especially lipids, in a number of rare diseases.

12

The Urinary System

The urinary system
  right and left kidneys
  right and left ureters
  bladder
  urethra

KIDNEYS

Each kidney (Fig. 12.1) is about 12 cm long, 7 cm wide and 2.5 cm thick at its maximum thickness, and lies at the back of the abdomen, posterior to the peritoneum, in the gutter that runs alongside the vertebral bodies. The perinephric fat is the fat in which they are embedded. The right kidney is placed slightly lower than the left because of the presence of the liver on the right side. An adrenal gland sits on the top of each kidney.

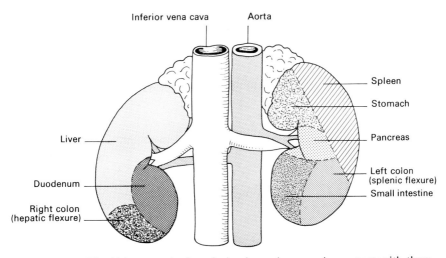

Fig. 12.1. The kidneys and adrenal glands, and organs in contact with them.

143

Each kidney has an upper and lower rounded end (the superior and inferior poles), a rounded convex lateral border, and on the medial border a depression called the hilum. The renal artery and vein, lymphatics, nerves, and the upper end of the ureter (Fig. 12.2) join the kidney at the hilum.

Blood supply: by the renal artery from the aorta; the right renal artery passes behind the inferior vena cava. The amount of blood passing through the kidneys is very great.

Venous drainage: by the renal vein into the inferior vena cava; the left renal vein passes in front of the aorta.

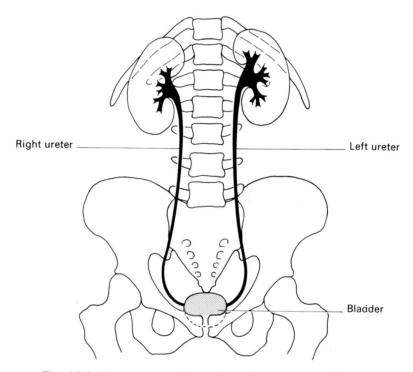

Right ureter _____ Left ureter

Bladder

Fig. 12.2. The ureters as shown by an intravenous pyelogram.

Structure

(a) a fibrous capsule on the outside,
(b) a cortex, pale and spotted with glomeruli,
(c) a medulla, dark and streaky and composed of a number of conical renal papillae, which project into the pelvis, the enlarged upper end of the ureter.

Nephron

A *nephron* (Fig. 12.3) is the structural and functional unit of the kidney. Each kidney is composed of about one million nephrons.

Each nephron is composed of a renal tubule, its glomerulus and associated blood vessels.

Each *renal tubule* is a long bent tube, lined by a single layer of cuboidal cells. It begins as *Bowman's capsule*, a double-layered cup enclosing the glomerulus; twists upon itself to form the proximal covoluted tubule; runs from the cortex to the medulla and back again, forming the *loop of Henle*; twists upon itself again to form the distal convoluted tubule; and ends by entering a collecting tubule.

Each collecting tubule runs through the medulla of the kidney, is joined by collecting tubules from other nephrons, and they open together upon the surface of a renal papilla within the pelvis of the ureter.

The *blood vessel* to the nephron runs a special course:

(a) the *glomerulus* is a whorl of capillaries enclosed within Bowman's capsule. An afferent arteriole brings blood to it.

(b) An efferent vessel passes from the glomerulus to the renal tubule and divides into capillaries on its surface.

(c) These capillaries drain into a vein, which joins other veins eventually to form the renal vein.

It is essential to appreciate that the blood from the glomerulus goes to the tubule of the same nephron before passing into a vein and leaving the kidney.

The kidneys are the main excretory organs of the body, and in order that they can carry out these excretory functions they have to receive a large proportion—about one-quarter when the body is at rest—of the blood pumped out of the heart with every beat.

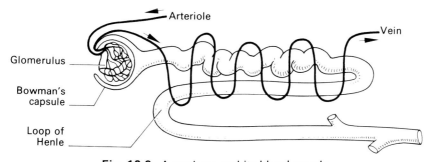

Fig. 12.3. A nephron and its blood supply.

> *Functions of kidneys*
> maintenance within normal limits of pH of
>   fluids of body, their amount and composition
>   and in particular of the amounts of sodium and
>   potassium in them
> excretion of some end-products of metabolism
> hormone secretion
> a role in production of vitamin D
> excretion of some drugs

Regulation of body fluids and control of acid–base balance

Different activities take place in the glomerulus and the tubule of a nephron.

Glomerular action

Under the pressure that is set up in the capillaries, a glomerular filtrate passes out of the blood in the glomerulus into the dilated end of the tubule. This filtrate is identical with blood plasma except that it does not contain certain plasma proteins and other molecules too large to pass into the tubule. In a healthy adult about 120 ml per minute are filtered out or about 170 litres in 24 hours. The amount declines with age.

Tubular action

As the filtrate passes along the various parts of the tubule and collecting duct, its amount and composition are altered by interchanges between it and the blood in the capillaries on the wall of the tubule and collecting duct. Of the 170 litres of glomerular filtrate in 24 hours about 147 litres are reabsorbed in the tubule. *Substances reabsorbed into the blood*:

    water (90 per cent)
    chloride (90 per cent)
    sodium (90 per cent)
    glucose (100 per cent)
    bicarbonate
    amino acids   calcium

Substances passing out of the blood into the filtrate in order to maintain constant the pH of the body fluids:

    hydrogen ions
    ammonium salts   phosphates
    potassium (in exchange for sodium)

The *antidiuretic hormone* (ADH), which is secreted by the hypothalamus and transferred thence to the posterior lobe of the pituitary gland, controls the further reabsorption of water by regulating the permeability of the collecting duct. Of the 23 litres of fluid passing into the ducts in 24 hours over 20 are by this action reabsorbed.

By all these changes the fluid has now become urine, of which 1–1.5 litres are excreted in 24 hours.

In normal conditions all the nephrons of a kidney are not working at the same time or not working to full capacity. Some are in a resting state, forming a 'renal reserve', which can be brought into action should a need arise for greater excretion.

The excretion of the end-products of metabolism

Urea, uric acid and creatinine are excreted in the urine as end-products of protein metabolism.

The production of hormones

See p. 220.

The production of vitamin D

The kidney is believed to take the final action in the production of a biologically active vitamin D (see p. 193).

The excretion of drugs

The excretion of drugs is mainly through the kidneys. The concentration of some drugs in them can be high and can have a toxic effect on the cells of the tubules. Diseases of the kidney, by preventing the excretion of a drug, can 'potentiate' it (i.e. increase the effect of a dose) by retaining it in the body.

Urine
    amount: 900–1500 ml per 24 hours (varies with
        fluid intake and amount of fluid lost by other
        routes)
    specific gravity (SG): 1002–1030 (is indication
        of amount of substances dissolved in it)
    reaction: acid, pH about 6.0 (on ordinary diet)
    colour: due to urochrome (a pigment of
        uncertain origin)
    composition:
    water
    urea   20–30 g in 24 hours
    uric acid   0.6 g in 24 hours
    creatinine   1–2 g in 24 hours
    ammonia
    sodium   potassium   phosphates
    chloride   sulphates

Congenital abnormalities

Congenital abnormalities of the kidney can occur, including:
(a) absence of a kidney,
(b) a horseshoe kidney, the two kidneys being joined below by renal tissue,
(c) cystic kidney, in which the kidneys have a large number of cysts as a result of a developmental error in the development of the tubules.

Diseases of the kidney

Diseases of the kidney can interfere with the functioning of nephrons, and if a large number of nephrons are thrown out of action evidence of impaired renal function appears: the secretion of urine is diminished, albumin or blood may appear in the urine, the products of metabolism (e.g. urea) which should be excreted are not excreted and accumulate in the blood, and the acid–base balance of the body is disturbed.

In *acute glomerular nephritis* the kidneys are inflamed, the glomeruli being particularly affected. In the *nephrotic syndrome* a loss of protein in the urine leads to gross fluid retention in the tissue spaces. In *renal glycosuria* glucose leaks into the urine as the result of a congenital defect in the anatomy and functioning of nephrons.

Acute renal failure can arise as a result of (a) an impairment of renal circulation (e.g. in shock the reduced cardiac output is directed to the brain and heart to the detriment of the kidney); (b) severe glomerular nephritis; (c) a blocking of the urinary tract by stones. If renal failure persists for more than a few hours, the renal tubules suffer permanent damage. The amount of urine excreted is stopped completely (anuria) or reduced to a small amount (oliguria); there is a severe disturbance of the acid–base balance and the body's metabolic end-products are not excreted. *Chronic renal failure* is a result of permanent damage to nephrons by any severe kidney disease; evidence of renal failure begins to appear when about 75 per cent of the nephrons are knocked out of action.

In *diabetes insipidus* the antidiuretic hormone is not produced by the hypothal-amus–pituitary complex and in consequence water is not reabsorbed in the collecting ducts, and the patient passes large quantities of dilute urine.

Abnormal constituents of urine
  glucose
  ketone bodies
  bile salts
  bile pigments
  protein
  blood
  some drugs

URETER

The ureter (Fig. 12.2, p. 144) is a tube from the kidney to the bladder. Each
(a) is about 25 cm long,
(b) starts as a pelvis, a dilated part attached to the hilum of the kidney,
(c) runs down the posterior abdominal wall behind the peritoneum,
(d) in the pelvis turns forwards and inwards to enter the bladder, through whose wall it runs obliquely.

Structure

The ureter has a mucous membrane lined with cuboidal epithelium, and a thick muscular wall.

Urine is squeezed down the ureter by peristaltic waves, which occur about 1–4 times a minute; urine enters the bladder in a series of squirts. The oblique entry through the bladder wall ensures that the bottom end is closed during micturition by the contraction of the bladder, thus preventing reflux of urine up the ureter and preventing the spread of infection from the bladder upwards.

Clinical features

Hydronephrosis is dilatation of the pelvis of the ureter produced by an obstruction to the outflow of urine by a stone or by an abnormally placed artery pressing on the ureter; the pelvis enlarges and there is a progressive destruction of renal tissue. *Pyonephrosis* is pus in the pelvis. *Pyelitis* is an infection of the pelvis. The *passage of a renal stone* down the ureter causes intense pain in the loin.

BLADDER

The bladder is a muscular sac into which urine passes from the ureters. When empty or half distended it lies in the pelvis; when more than half distended it rises up in the abdomen above the pubis.

RELATIONS

in front: the symphysis pubis,

behind in the male (Fig. 12.4): the termination of the vasa deferentes, the seminal vesicle, the rectum,

behind in the female (Fig. 12.5): the uterus and vagina,

above: coils of small intestine, and in females the anterior end of the body of the uterus,

at the sides: the levator ani muscle, pelvic fascia and ligaments,

below in the male: the prostate gland,

below in the female: the anterior vaginal wall.

Peritoneum covers the upper surface of the bladder, rising up with it as the bladder distends, so that the distended bladder is immediately behind the anterior abdominal wall. The *rectovesical pouch* is the pouch of peritoneum between rectum and bladder in men. The *vesico-uterine pouch* is the pouch of peritoneum between bladder and uterus in women.

The ureters enter the bladder posteriorly, their opening being about 5 cm apart. The urethra leaves the bladder inferiorly.

The *trigone* is the triangular area of smooth mucous membrane between the openings of the ureters and urethra.

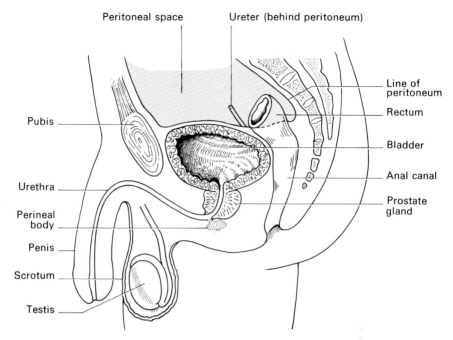

Fig. 12.4. Section through the male pelvis

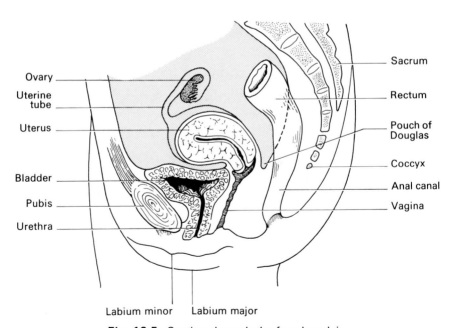

Fig. 12.5. Section through the female pelvis.

Structure

The bladder is composed of:
(a) a mucous membrane: thrown into folds when the bladder is empty,
(b) a submucous coat,
(c) a muscular coat: forms the greater part of the thickness of the bladder wall; part of it forms a sphincter around the opening of the urethra,
(d) peritoneum or pelvic fascia on the outside.

Micturition

Micturition is essentially a reflex action which after infancy is controlled by centres higher in the nervous system.

The entry of urine into the bladder in time starts to stretch the muscle fibres of the bladder wall. Impulses pass along afferent nerves to the lumbar part of the spinal cord, and being transmitted to the cerebral cortex produce a desire to micturate. The degree of stretching necessary to produce this effect varies with individuals, some being able to tolerate much distension without discomfort.

Micturition is controlled through the efferent nerves to the bladder. Impulses passing along the sacral parasympathetic nerves cause:

the muscle of the bladder wall to contract,

the sphincter of the bladder to relax.

The expulsion of urine is aided by contraction of the muscles of the abdominal wall and of the diaphragm, which produces a collapse of the bladder by raising the intra-abdominal pressure.

Clinical features

The bladder becomes grossly distended if there is an obstruction to the outflow of urine through the urethra, e.g. in men by an enlargement of the prostate gland. It is possible to drain a full bladder or obtain a specimen of urine by a suprapubic stab, a needle being inserted through the abdominal wall directly into the bladder; there should be no danger of puncturing the peritoneum as this will have risen up with the top of the bladder out of the way. *Cystitis* is an inflammation of the bladder; the urine is particularly likely to become infected if there is an obstruction to its outflow. *Cystoscopy* is the examination of the interior of the bladder through a cystoscope, a tube with an arrangement of mirrors and lights, which is passed up the urethra; it is fitted with thin tubes which can be inserted into the openings of the ureters in order that urine from each kidney can be separately collected.

MALE URETHRA

The male urethra is a tube about 20 cm long and extends from the bladder to the end of the penis. It has three parts:

prostatic urethra: is 3 cm long; traverses the prostate gland; receives the two ejaculatory ducts and several small ducts from the prostate gland.

membranous urethra: is 2 cm long; traverses the urogenital diaphragm, a fibrous sheet just below the prostate gland; is enclosed by a sphincter of muscle fibres. It is called membranous because here it is as thin as a membrane.

spongy urethra: is about 15 cm long; passes through the corpus spongiosum of the penis to end on its tip.

Urine is forced along the urethra by contraction of the bladder. The last drops in the urethra are expelled by contraction of the sphincter encircling the membranous urethra.

Clinical features

Rupture of the urethra can follow a fall astride; if the patient then empties his bladder, the urine passes into the soft tissues around the urethra. *Urethritis* is inflammation of the urethra; the common causes are gonorrhoea and non-specific venereal infection. *Stenosis* (narrowing) of the urethra is due to scar tissue forming after the healing of a rupture or an attack of gonorrhoea.

FEMALE URETHRA

The female urethra is a tube about 3 cm long and extends from the bladder to an opening between the labia minora about 2.5 cm behind the clitoris. It runs immediately in front of the vagina.

Clinical feature

Urethritis, inflammation of the urethra, is most commonly due to gonorrhoea and non-specific venereal infection.

13
Fluids and Electrolytes in the Body

For health and life fluids and electrolytes have to be present in the right proportions in various tissues. This is achieved by a series of complex physico-chemical manoeuvres.

Terms

acid: any substance that gives up hydrogen ions

acidaemia (*acidosis*): a fall in pH below the normal

alkalaemia (*alkalosis*): a rise in pH above the normal

anion: an ion with a negative electric charge (bicarbonate, chloride, sulphate, phosphate, protein etc)

base: any substance that received hydrogen (H) ions (bicarbonate, dibasic phosphate, some amino acids)

buffer solution: one which resists a change in pH when small amounts of acid or base are added

cation: an ion with a positive electric charge (hydrogen, sodium, potassium etc)

electrolyte: any acid or base etc, which when dissolved produces ions

ion: an electrically charged particle

pH: an indication of the concentration of hydrogen (H) ions in a solution and hence of the degree of acidity.

WATER IN THE BODY

There are about 50 litres of water in the body of an average man weighing 70 kg
water forms 75 per cent of the body of an infant,
water forms 70 per cent of the body of an adult man,
water forms 55 per cent of the body of an old man.
Because women have relatively more fat (which is relatively water-free), the amount of water in a woman is about 10 per cent less that that in a man.
The water is contained in two main 'compartments' in the body.

Intracellular fluid (ICF): i.e. the water inside the various cells of the body. This is about 70 per cent of the total amount of water in the body.

Extracellular fluid (ECF): i.e. the water not in the cells. This is about 30 per cent of the total amount of water in the body. It is present in:

interstitial fluid: this is found in the tissue spaces between cells,

plasma of the blood,

cerebrospinal fluid, lymph, fluid in serous cavities and joints: in amounts too small to be important in water balance.

ICF is the medium in which the chemical activities of cells take place. ECF is the medium for the transport of chemical substances from one cell to another.

A series of complicated exchanges takes place in both directions between ICF and ECF in order to maintain a correct chemical and electrolyte balance and to maintain a normal pH.

Electrolytes			
Anions		*Cations*	
Chloride	Cl	Sodium	Na
Sulphate	SO_4	Potassium	K
Phosphate	HPO_4	Calcium	Ca
Bicarbonate	HCO_3	Magnesium	Mg

Water balance

Water balance (Fig. 13.1) is achieved by an equal gain and loss of water. As water is being inevitably lost all the time through the kidneys, the skin and the lungs, the main problem is to keep enough water in the body.

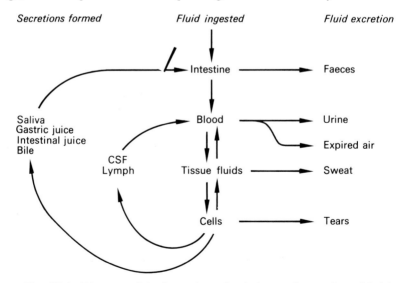

Fig. 13.1. Diagram of the ingestion, circulation and excretion of fluid.

GAIN OF WATER

Water is gained in three ways:
(a) by drinking: the amount gained is basically controlled by thirst, but is much affected by drinking habits. Thirst is not always due to dryness of the mouth; in the hypothalamus of the brain there may be a 'drinking centre' which reacts to dehydration.
(b) by eating: food contains water,
(c) by the oxidation of food in the body.

LOSS OF WATER

Water is lost in four ways:
(a) as urine: about 1.5 litres daily, varying with water intake and loss of water by other routes, e.g. by sweating,
(b) in expired air from the lungs: about 400 ml daily,
(c) in faeces: about 100 ml daily,
(d) by the skin: as sweat or invisible perspiration, the amount varying with the temperature, humidity and circulation of air, the amount of clothing worn, the amount of work done.

The main water loss—that through the kidney—is partly inevitable, partly controlled by the antidiuretic hormone (ADH). ADH is produced in the hypothalamus and transported to the pituitary gland, whence it is discharged as required. It regulates the reabsorption of water from the distal tubules of the kidney and so regulates the amount of urine excreted.

pH

The pH is an indication of the concentration of hydrogen (H) ions in a solution and hence of the degree of acidity. *The normal pH of body fluids is: 7.36-7.44.*

There is confusion in the use of the terms acidaemia, acidosis, alkalaemia, alkalosis. *Acidaemia* is used for severe falls in pH, i.e. below 7.3. *Alkalaemia* is used for rises above 7.5. Acidosis and alkalosis are sometimes used as if they were interchangeable with acidaemia and alkalaemia, sometimes to describe minor and easily compensated changes in pH.

For normal metabolic processes to take place, the pH must remain within normal limits. *Buffers* in the blood help to maintain this stability. A buffer solution is one which resists change to pH by being able to absorb small amounts of acid or base.

Carbon dioxide and acids produced by metabolic processes tend to lower the pH and have to be excreted. Carbon dioxide is excreted by the lungs in expired air and the acids by the tubules of the kidneys. If the amount of base is increased, respiration is reduced in rate and depth, with the result that less carbon dioxide is excreted; the kidneys react by excreting bicarbonate and conserving H ions.

Buffers
in plasma
    bicarbonate (the principal buffer base)
    protein
    phosphate

in red cells
    haemoglobin

in urine
    phosphates

Clinical features
Disorders of fluid and electrolyte balance can occur in many conditions, with the production of:
    fluid deficiency or excess,
    sodium deficiency or excess,
    potassium deficiency or excess.
Fluid deficiency: occurs in coma, with an inability to swallow, having no water to drink as in a desert or in a boat at sea.
Fluid excess produces *oedema*, an excess of fluid in the tissue spaces. It occurs in heart failure, the nephrotic syndrome (a kidney disease), obstruction to the lymph drainage of a part of the body, in famine etc. The fluid tracks down in the tissue spaces to the lowest parts of the body.
Sodium deficiency occurs in profuse sweating if fluid only is replaced and not salt, and in Addison's disease, a disease of the adrenal cortex.
Fluid and sodium deficiency occurs in severe vomiting and severe diarrhoea.
Potassium deficiency occurs in diarrhoea, diabetic crisis, prolonged treatment with diuretic drugs, when a patient has been given prolonged glucose-saline infusion without added potassium.
Potassium excess occurs in severe tissue injuries and chronic renal failure.

Causes of acidaemia
    diabetes mellitus: excess acid production in
        diabetic crisis
    chronic renal failure: reduced acid excretion
    hypoventilation (reduced breathing): in
        morphine poisoning, respiratory tract
        obstruction, respiratory muscle paralysis

Causes of alkalaemia
    persistent vomiting
    excessive consumption of sodium bicarbonate in
        treatment of peptic ulcer
    hyperventilation (over-breathing): in salicylate
        poisoning and some hysterics

14

The Alimentary System

THE MOUTH

The boundaries of the mouth (Fig. 14.1, 14.2) are:
    above: the hard and soft palate,
    below: the mandible, tongue and other structures in the floor of the mouth,
    laterally: the cheeks,
    in front: the lips,
    behind: an opening into the pharynx.
The *cheek* is formed by:
    mucous membrane,
    buccinator muscle, which extends from the maxilla to the mandible,
    buccinator pad of fat (which, well developed in infancy, gives the baby his chubby appearance),
    skin.

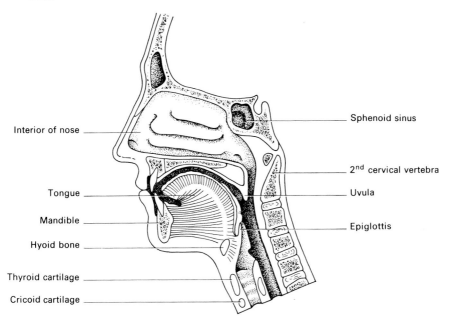

Fig. 14.1. Section through the mouth, nose and pharynx.

157

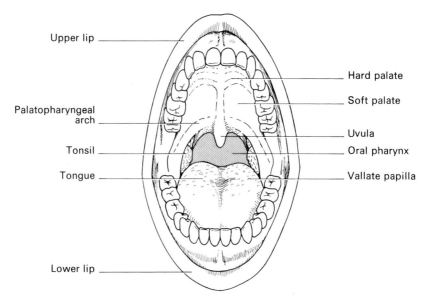

Upper lip

Palatopharyngeal
arch

Tonsil

Tongue

Lower lip

Hard palate

Soft palate

Uvula

Oral pharynx

Vallate papilla

Fig. 14.2. The interior of the mouth.

The *floor of the mouth* is formed by:
>the tongue,
>a depression in front of and at the sides of the tongue where mucous
>membrane is reflected from tongue to gums,
>beneath this depression, the submandibular and sublingual salivary glands
>and some small muscles acting on the tongue.

The *hard palate* is formed by parts of the maxilla in front and the palatine
bone behind. The bones are covered by periosteum and mucous membrane. The
soft palate, composed of muscular and fibrous tissue enclosed in mucous membrane, is continuous with the hard palate in front. The *uvula* is a small soft
conical process hanging down in the midline. On either side are two arches of
mucous membrane between which is the tonsil.

The tongue

The tongue is composed of muscle enclosed at the top and sides by mucous
membrane.

The *dorsum* forms part of the floor of the mouth and is curved backwards
and downwards, its posterior third facing the pharynx and not normally visible.
The *sulcus terminalis* is a V-shaped groove, with the V pointing backwards,
which separates the anterior two-thirds from the posterior third. The *foramen
caecum* is a small pit at the apex of the V. The mucous membrane of the dorsum
is thick and covered with numerous papillae. About 12 large papillae are visible
in a row just in front of the sulcus terminals; each is surrounded by shallow
ditch. The *taste-buds* are specialized cells in the walls of these ditches and

contain the cells in which tastes are appreciated and from which they are communicated to the brain. The *root*, the postero-inferior part of the tongue, is attached by muscles to the palate, styloid process of the temporal bone, the mandible and the hyoid bone. The *frenulum* is a short fold of mucous membrane in the midline passing from just below and behind the tip of the tongue to the floor of the mouth.

Blood supply: lingual artery (a branch of the external carotid artery).

Lymph drainage: to cervical lymph nodes.

Nerve supply:

(a) *sensory*: lingual nerve (a branch of the mandibular, a branch of V cranial) serves anterior two-thirds for touch; facial nerve (VII cranial) serves anterior two-thirds for taste; glossopharyngeal nerve (IX cranial) serves posterior third for touch and taste,

(b) *motor*: hypoglossal nerve (XII cranial).

The salivary glands

The salivary glands are composed of saliva-secreting cells.

> *Salivary glands*
>   parotid, right and left
>   submandibular, right and left
>   sublingual, right and left

PAROTID GLAND

The parotid gland is a roughly wedge-shaped gland lying in front of, below and behind the pinna of the ear. The *parotid duct* arises from the anterior border, runs horizontally across the cheek, pierces the buccinator fat and muscle, opens inside the cheek opposite the 2nd upper molar tooth.

Branches of the facial nerve (VII cranial) pass forwards through the gland to reach the muscles of the face.

SUBMANDIBULAR GLAND

The submandibular gland lies at the back of the floor of the mouth under cover of the angle of the mandible. Its duct runs forward in the floor of the mouth to open into the mouth at the side of the tongue.

SUBLINGUAL GLAND

The sublingual gland lies under the mucous membrane of the floor of the mouth and under cover of the front of the tongue. It has about 12 small ducts which open into the floor of the mouth.

The salivary glands secrete saliva as a response to the anticipation of food or the presence of food in the mouth. Stimulation through parasympathetic nerves produces dilatation of blood vessels in the glands and a flow of saliva.

Saliva has three functions:

(a) It enables food to be masticated by the teeth and formed into a bolus, a lump suitable for swallowing,

(b) Ptyalin, an enzyme in saliva converts starch into maltose.

(c) By keeping the tongue and inside of the mouth moist, it enables the tongue to be moved in speech.

Clinical features

Lack of saliva causes a dry mouth, e.g. as in dehydration.

Parotitis is inflammation of the parotid gland; it occurs in mumps and as an acute infection in weak, debilitated, ill patients. *Stones* can form in the salivary ducts and block them.

The pharynx

The pharynx is a fibromuscular tube attached to the base of the skull above and continuous with the oesophagus below. It is composed of three parts, of which the nasopharynx and oropharynx have been described on p. 118. The laryngeal part is behind the epiglottis and larynx and continuous with the oesophagus below. Food has to pass through the oropharynx and laryngeal pharynx to enter the oesophagus.

Mastication

Food is cut and crushed by the teeth and moistened with saliva to form a bolus, a saliva-coated mass.

Swallowing
1. the bolus is squeezed by the tongue against the hard palate
2. the soft palate closes off the nasopharynx
3. the bolus descends down the dorsal surface of the tongue
4. at this stage the larynx is raised, the entrance to the larynx narrows, the vocal cords close
5. the position of the epiglottis directs the food away from the laryngeal opening, and after the bolus has passed it, the epiglottis bends over the laryngeal opening, possibly to prevent any residual crumbs from getting into the larynx
6. after the bolus has passed into the oesophagus, the upper structures resume their usual position

Oesophagus

The oesophagus (Fig. 14.3) is a muscular tube about 25 cm long and 0.5 cm in diameter.

It begins in the neck as a continuation of the pharynx, passes down the neck and thorax and then through the left crus of the diaphragm to enter the stomach.

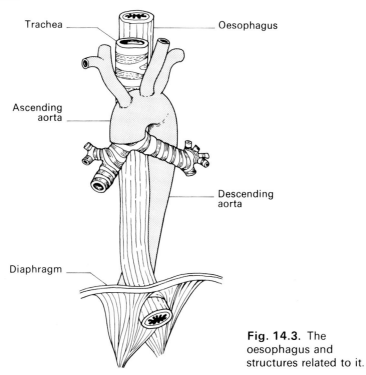

Trachea — Oesophagus

Ascending aorta

Descending aorta

Diaphragm

Fig. 14.3. The oesophagus and structures related to it.

In *front of it* are:
    trachea and thyroid gland,
    heart,
    diaphragm.
Behind it is:
    the vertebral column.
On either side are:
    lungs and pleurae.
The arch of the aorta lies on its left side and the descending aorta lies at first on the left and then passes behind it, coming to lie between the oesophagus and the vertebral column.

The oesophagus is slightly narrow:
(a) at its upper end,
(b) where the left bronchus crosses it,
(c) where it passes through the diaphragm.

COMPOSITION
(a) inner layer of mucous membrane,
(b) thick submucous coat, containing mucous glands,
(c) muscular coat of longitudinal and circular fibres,
(d) outer fibrous coat.

The bolus falls down the upper third of the oesophagus in a fraction of a second and is pushed down the rest by a ring-like contraction of oesophageal muscle. A moist soft bolus reaches the entrance to the stomach in a few seconds, but a dry bolus may have to be pushed down by secondary waves, which can be painful.

Clinical features
The passage of food down the oesophagus can be prevented by:
    stenosis, a narrowing of its lumen due to scar tissue following scalding,
    carcinoma of the oesophagus,
    achalasia, a condition in which the normal reflex relaxation at the bottom end of the oesophagus does not occur and peristalic waves higher up are disorganized.
    Swallowed foreign bodies, such as small toys, are liable to lodge at one of the three narrow places.

THE STOMACH

The stomach (Figs 14.4, 14.5, 14.6) is a wide and very dilatable part of the alimentary tract. It varies in shape according to the amount of food in it, the presence of peristaltic waves, pressure from other organs, respiration, and the posture of the body. The position, shape and mobility of the stomach are variable.
    It usually has a J-shaped form and lies in the upper left quadrant of the abdomen.

It shows:
    anterior and posterior surfaces,
    a lesser curvature on the right side,
    a greater curvature on the left side,
    a cardiac orifice where the oesophagus joins it,
    a fundus: the dome above the level of the cardiac orifice, normally occupied by a bubble of air,
    a body: the greater part of the stomach,
    a pyloric canal: a narrow tube below the body,
    a pyloric opening: into the first part of the duodenum.
    The *cardiac orifice* has no special sphincter (ring of muscle opening and closing it by its contraction and relaxation), but is kept closed by folds of mucous membrane and by muscle fibres at the bottom of the oesophagus.

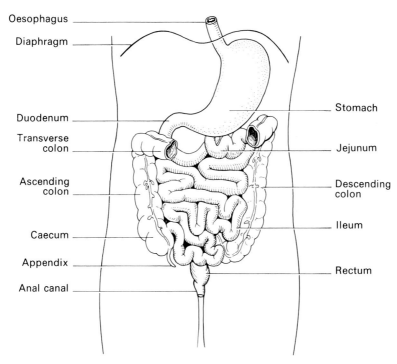

Fig. 14.4. The gastrointestinal tract. Part of the transverse colon has been removed.

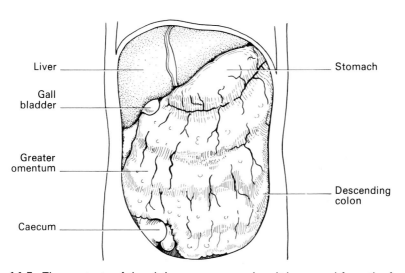

Fig. 14.5. The contents of the abdomen as seen when it is opened from the front.

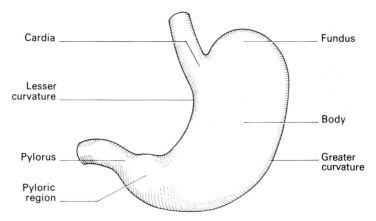

Fig. 14.6. The stomach

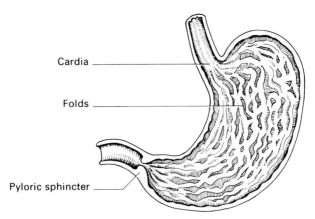

Fig. 14.7. The stomach from within showing folds of mucous membrane and the pyloric sphincter.

The *pyloric opening* is surrounded by the pyloric sphincter (Fig. 14.7), a definite sphincter formed by a thickening of the circular muscles of the stomach.

Peritoneal covering. The stomach is covered by peritoneum, except along the lines of the greater and lesser curvatures, off which it is reflected in double folds.

The greater curvature is connected to the transverse colon by the greater omentum, a double fold of peritoneum, and to the spleen.

The lesser curvature is connected to the liver by the lesser omentum, a double fold of peritoneum.

Structure

(a) mucous membrane: vascular, red, thrown into folds, and showing the openings of the ducts of millions of glands of several kinds,

(b) submucous coat: of loose areolar tissue,

(c) muscular coat: of circular, oblique and longitudinal muscle fibres,
(d) peritoneal coat.

Blood supply: by gastric arteries, from the coeliac artery and its branches; run along the greater and lesser curvatures between the folds of peritoneum and give branches to both sides of the stomach.

Venous drainage: into the portal system.

Lymph drainage: into nodes along its two curvatures, around its two openings; thence into aortic nodes and the cisterna chyli.

Nerve supply: by sympathetic and parasympathetic (vagal) nerves.

> *Functions of stomach*
>   to act as a 'hopper', containing food in a sac and
>   releasing it gradually into the intestine
>   to continue with the digestion of food into such a
>   state that further digestion can take place in
>   the intestine
>   to secrete the 'intrinsic factor' necessary for the
>   absorption of vitamin B_{12}

GASTRIC JUICE

Gastric juice is a watery fluid secreted by the glands and cells of the mucous membrane of the stomach. It contains:
  hydrochloric acid in dilute solution,
  pepsinogen: which is converted by the acid in the stomach into *pepsin*; pepsin splits proteins into smaller molecules,
  mucus: secreted from cells on the surface of the mucous membrane; its principal function is to coat the surface of the mucous membrane to protect it from digestion by hydrochloric acid.

Secretion of gastric juice
Gastric juice is secreted in three phases.
  Cerebral phase. The anticipation of food causes stimuli to pass from the brain down the vagus nerves to the stomach where the glands and cells are stimulated into secretion. In this phase *gastrin*, a hormone secreted by cells of the mucous membrane of the pyloric canal of the stomach, enters the bloodstream and eventually arrives back in the mucous membrane of the stomach which it stimulates to produce more gastric juice.
  Gastric phase. More gastrin is produced by a combination of three events— the mechanical stretching of the stomach by food, the presence of protein products in the stomach, and vagal stimulation.

Intestinal phase. The arrival of food in the small intestine causes a further secretion of gastric juice, possibly by the production of more gastrin.

Digestion in the stomach

The amount of digestion carried out in the stomach is small, being limited to the conversion of proteins into peptones.

Achlorhydria is absence of hydrochloric acid from the gastric juice. It occurs in 1 per cent of people. It is associated with a failure of secretion of the intrinsic factor needed for the absorption of vitamin B_{12}.

MOVEMENTS OF THE STOMACH

In the resting state the stomach is contracted. If the next meal is not forthcoming, it develops peristaltic waves, which cause pangs of hunger.

The stomach distends to accommodate the food that arrives, and then peristaltic waves begin in its upper part and pass down towards the pylorus, as many as four being present at one time. At first the pylorus remains closed, and the effects of the waves at this time are to mix up the food and expose it to gastric juice. Then the pyloric sphincter starts to relax and let small amounts of food through at a time.

Emptying can take up to 5 hours. It can be longer if the person is worried or if there is a lot of fat in the food.

Clinical features

A *barium meal*, a paste containing barium sulphate, which is opaque to X-rays, is swallowed in order to show the size, shape, peristaltic waves, time of emptying, and the presence of an ulcer or cancer.

Gastroscopy is inspection of the mucous membrane of the stomach through a gastroscope, a tube with lenses and a light, which is passed through the mouth, pharynx and oesophagus into the stomach.

Gastrin secretion test is a method of obtaining gastric juice for analysis. A Ryle's tube, a thin rubber tube, is passed into the stomach; an injection of pentagastrin (which produces the same effect as gastrin) is given and the gastric juice secreted is aspirated up the tube at intervals and examined for its content of hydrochloric acid and for the presence of blood or cancer cells.

Gastritis is inflammation of the stomach. A *gastric ulcer* is an erosion of the stomach wall, usually in the lower one-third along the lesser curvature. A similar ulceration can occur in the duodenum, producing a duodenal ulcer. *Peptic ulcer* means a gastric or duodenal ulcer. The ulcer may be superficial, involving only the mucous membrane, or may penetrate the other layers of the stomach wall. The cause is unknown. One factor may be a local lack of mucus, so that the hydrochloric acid in the stomach is not prevented from digesting the mucous membrane. *Cancer of the stomach* causes loss of appetite and weight, and is usually fatal.

THE SMALL INTESTINE

The small intestine is the part of the alimentary tract between the stomach and the large intestine. It is a long, much coiled tube which fills the greater part of the abdominal cavity.

Parts of small intestine
duodenum
jejunum
ileum

Duodenum

The duodenum (Fig. 14.8) is a C-shaped tube, about 25 cm long, at the back of the abdomen, curved round the head of the pancreas. It is described in 4 parts,

1st part: running to the right,

2nd part: running downwards,

3rd part: running transversely to the left and in front of the inferior vena cava and aorta,

4th part: running upwards to become continuous with the jejunum.

The stomach opens into the 1st part at the pyloric opening.

The pancreatic and bile ducts open into the 2nd part by a common opening on a small papilla, the opening being controlled by a sphincter called the sphincter of Oddi. Sometimes the ducts open separately.

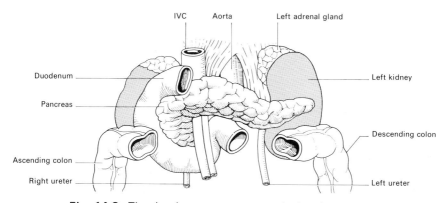

Fig. 14.8. The duodenum, pancreas and related structures.

Jejunum and ileum

The jejunum is the first part and the ileum the second part of the rest of the small intestine. Their combined length varies from 300 to 900 cm.

Their names are traditional. There is no sharp distinction between them. The jejunum is slightly bigger, has a thicker wall, more folds of mucous membrane and fewer Peyer's patches.

They are enclosed within peritoneum except along the line of its attachment.

Blood supply of small intestine: by branches of the superior mesenteric artery (a branch of the aorta); the branches are linked in the mesentery by a number of arcades of arteries, from which arise the terminal branches.

Venous drainage of small intestine: into the superior mesenteric vein and so into the portal vein.

Lymph drainage of small intestine: into nodes in the mesentery and thence into aortic glands and the cisterna chyli.

Nerve supply of small intestine: by sympathetic and parasympathetic (vagal) nerves.

Functions of small intestine
1. secretion of intestinal juice
2. reception of bile and pancreatic juice
3. digestion of foodstuffs
   Intestinal juice and pancreatic juice contain enzymes which convert:
   proteins into amino acids
   carbohydrates into glucose, maltose and galactose
   fats into fatty acids and glycerol (with the help of bile salts in bile discharged into the duodenum by contraction of gall-bladder).
   Digestion is completed, the foodstuffs being broken down into simple forms which are absorbed through the wall of the small intestine into the blood or lymph.
4. absorption of water, salts and vitamins
5. movement of intestinal contents along the intestine by short segmental contractions and 'rush waves' which move the contents along more quickly.

STRUCTURE

Mucous membrane: this is thrown into numerous circular or semicircular or spiral folds. The whole surface is marked by millions of villi; a *villus* is a tiny projection covered with a single layer of cells and containing blood vessels, lymph vessels, nerves and muscle fibres.

Peyer's patches are patches of lymph tissue in the mucous membrane; they are more common in the ileum than the jejunum.

submucous coat.

muscular coat: of circular and longitudinal fibres.

peritoneum.

Meckel's diverticulum: a congenital abnormality found in 2 per cent of people. It is a diverticulum of the small intestine up to 5 cm long, occurring at 15–60 cm from the ileocaecal valve. It is a remnant of the duct which connected the yolk sac to the intestine and may be connected to the umbilicus by a fibrous cord, which is a remnant of the duct.

Digestion

The further digestion of food is completed in the small intestine by the action of pancreatic juice, bile and intestinal juice.

Clinical features

A *duodenal ulcer* (similar to a gastric ulcer) can form in the 1st part of the duodenum, where the partly digested food is acid. Below the level of the pancreatic–biliary opening, the duodenal contents are made alkaline by pancreatic juice and bile and ulceration does not occur.

In the *malabsorption diseases* (e.g. idiopathic steatorrhoea) there is a failure of the small intestine to absorb foodstuffs, vitamins and salts. This causes wasting, vitamin deficiency, salt deficiencies and anaemia.

Meckel's diverticulum can become inflamed. In *typhoid fever* Peyer's patches become inflamed and can ulcerate, sometimes producing a perforation into the peritoneal cavity.

THE LARGE INTESTINE

The length of the large intestine is variable, averaging about 150 cm. It is distinguishable from the small intestine by its larger size and the presence of taeniae coli and appendices epiploicae. The *taeniae coli* are 3 bands of longitudinal muscle fibres on the outside of the colon and being shorter than the rest of the bowel wall give it a sacculated or puckered appearance. The appendix and rectum do not show taeniae coli. The *appendices epiploicae* are tags of fat-containing peritoneum on the surface of the caecum.

Large intestine
caecum
appendix
ascending colon
transverse colon
descending colon
sigmoid (pelvic) colon
rectum
anal canal

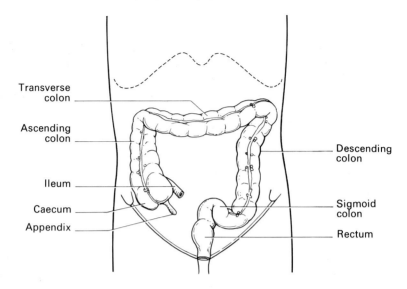

Fig. 14.9. The colon

Caecum

The caecum (Fig. 14.10) is a wide sac, lying in the right iliac fossa. The ileum enters its left side at the ileocaecal opening, an oval slit controlled by a muscular sphincter. The appendix opens into it below the ileocaecal opening. The caecum is continuous above with the ascending colon.

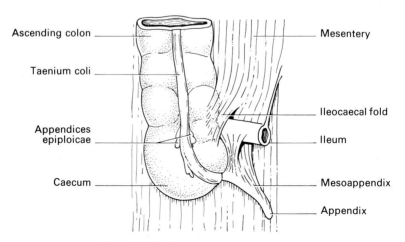

Fig. 14.10. The caecum and appendix.

Appendix

The appendix is a worm-like diverticulum up to 18 cm long and opens out of the caecum about 2.5 cm below the ileocaecal valve. It has a narrow lumen. Its submucous coat contains much lymph tissue.

It is connected to the mesentery of the ileum by a short triangular mesentery in which run the appendicular blood vessels and lymph vessels.

Its position is variable. In order of frequency it lies
behind the caecum,
below the caecum or hanging into the pelvis,
in front of or behind the end of the ileum,
in front of the caecum.

Clinical feature
Appendicitis is inflammation of the appendix. The cause is usually unknown, but it often follows a blocking of the lumen. In acute appendicitis the appendix becomes red and swollen, can go on to become gangrenous, or can ulcerate and produce peritonitis or an appendix abscess.

Ascending colon

The ascending colon extends from the caecum in the right iliac fossa up the right side of the abdomen to the right colic flexure under the right lobe of the liver.

Transverse colon

At the right colic flexure the colon bends sharply to the left and passes at the transverse colon across the abdomen in a loop which may hang lower than the umbilicus, and then ascends on the left side to end in the left colic flexure under the spleen.

Descending colon

At the left colic flexure the colon bends again to pass downwards on the left side of the abdomen down to the brim of the pelvis, where it is continued as the sigmoid colon.

Sigmoid (pelvic) colon

The sigmoid colon has several loops within the pelvis and ends opposite the middle of the sacrum where it is continuous with the rectum.

Rectum

This is about 12 cm long and gets its name from being straight—or nearly so. It begins at the middle of the sacrum and ends at the anal canal.

Relations of rectum
posterior: lower half of sacrum and the coccyx,
lateral: levator ani muscle,
anterior: (a) men—bladder, seminal vesicles, prostate gland,
                (b) women—cervix uteri, vagina.

Anal canal
The anal canal is about 3 cm long, runs downwards and backwards, and ends at
the anus.
    On either side are the ischiorectal fossae.
    It has an internal and an external sphincter muscle, which control the
opening and shutting of the anus.

Blood supply of the large intestine: by branches of the superior mesenteric artery
as far as the left colic flexure and by branches of the inferior mesenteric artery.

Venous drainage of the large intestine: into superior and inferior mesenteric
veins, and so into the portal vein.

Lymph drainage of the large intestine: into lymph nodes in the peritoneum and
so into the aortic glands.

Nerve supply of the large intestine: by sympathetic and parasympathetic nerves.

STRUCTURE OF LARGE INTESTINE

(a) Mucous membrane: there are no circular folds, villi or lymph patches as in
the small intestine.
(b) submucous coat.
(c) Muscular coat: outer longitudinal fibres arranged in 3 taeniae coli; inner
circular fibres.
(d) Peritoneum: peritoneum covers the front and sides of the ascending and
descending colon; the transverse colon has a mesentery called the transverse
mesocolon; the sigmoid colon has a mesentery.

Functions of colon
    absorption from colonic contents of:
    water (2 or more litres in 24 hours)
    sodium
    chloride
    excretion of potassium into colonic contents

Movements of colon. The caecum and ascending and transverse colon fill up
gradually without any apparent movement of the colonic wall, but two or three
times a day the intestinal contents are moved along by powerful contractions.

Defecation

Defecation is in part a reflex, in part a voluntary activity. The entry of food into the rectum stimulates a desire to defecate, transmitted along afferent parasympathetic nerves to the sacral part of the spinal cord, and efferent messages are transmitted along efferent parasympathetic nerves to achieve muscular action. As a result of combined reflex and voluntary efforts:

1. relaxation of the sphincters of the anus occur,
2. colonic muscle contracts,
3. abdominal muscles and diaphragm contract,
4. pelvic floor is raised,
5. defecation occurs,
6. sphincters contract, expelling the last of the faeces.

Clinical features

Colitis is inflammation of the colon. *Ulcerative colitis* is a severe familial disease in which large areas of colon become ulcerated. *Cancer* can form in the large intestine—either as soft bleedings tumours or as slowly growing encircling rings of cancer, liable to produce intestinal obstruction.

Haemorrhoids (piles) are dilated blood vessels at the anal orifice.

Special tests of colonic function
examination of faeces
sigmoidoscopy—the visual examination of the
colonic mucous membrane through a
sigmoidoscope
proctoscopy—the visual examination of the
mucous membrane of the rectum through a
proctoscope
a barium enema and X-ray examination

PANCREAS

The pancreas is a long organ at the back of the upper part of the abdomen. It consists of a head (within the curve of the duodenum), neck, body and tail (which reaches the spleen). It is composed of (a) cells which secrete pancreatic juice and (b) intra-alveoli cell islets, also called islets of Langerhans (see. p. 219). The juice passes into a duct which traverses the length of the gland to join, in the head of the gland, the bile duct, the ducts opening together into the duodenum. The pancreatic juice is a digestive juice (see p. 168).

Clinical features

Pancreatitis is an acute or chronic inflammatory condition in which areas of pancreatic tissue are destroyed.

PERITONEUM

The peritoneum is a thin, smooth, moist serous membrane which lines the abdominal cavity and encloses many of the abdominal organs. The *parietal peritoneum* is the part lining the abdominal wall; the *visceral peritoneum* is the part which encloses an organ.

Essentially the peritoneum is a bag into which the organs have grown, bringing with them their blood vessels, lymph vessels and nerves.

The basic pattern has become modified by the growth of various organs and the way in which they have moved in fetal life into different positions.

Important features are:

The *parietal peritoneum* is continuous with the visceral peritoneum.

The *mesentery* is a double fold of peritoneum which attaches the jejunum and ileum to the posterior abdominal wall.

The *greater omentum* is a quadruple fold of peritoneum hanging down from the greater curvature of the stomach and covering the abdominal organs anteriorly like an apron; the transverse colon is enclosed within its two posterior layers.

The *lesser omentum* is a double layer of peritoneum connecting the lesser curvature of the stomach to the liver.

The *lesser sac* is the part of the peritoneal cavity that lies behind the stomach.

The *peritoneal cavity* is a 'potential' cavity, i.e. one which is squashed flat by the abdominal organs so that visceral and parietal peritoneum are touching. The cavity can be filled up by fluid or air in certain circumstances.

In women each uterine (fallopian) tube has at its free end a small opening into the peritoneal cavity for the entrance of the ovum. In men the peritoneal cavity is completely enclosed.

Functions of peritoneum
    attachment of organs to posterior abdominal wall
       and to one another
    organs enable to move over one another
    vessels and nerves allowed to reach organs
       without being twisted or squeezed
    store of fat
    sealing off of infected areas by great omentum

In the pelvis the peritoneum is thick and reflected off the pelvic organs.

The recto uterine *pouch* (pouch of Douglas) is made by the reflection of the peritoneum from the uterus on to the rectum.

Much *fat* is stored outside the peritoneum, especially round the kidney as the perinephric fat, and between its layers, especially in the omentum.

> *Structures lying wholly*
> *or mostly behind the*
> *peritoneum*
>
> | | |
> |---|---|
> | aorta | inferior vena cava |
> | cisterna chyli | pancreas |
> | duodenum | adrenal glands |
> | kidneys | ureters |

Clinical features

Peritonitis is an infection of the peritoneum produced by infected material being introduced into the peritoneal cavity, e.g. by a stab wound, the perforation of a peptic ulcer, the perforation of an inflamed appendix. The greater omentum can move and seal off a small infected area and prevent the infection from spreading to other parts of the peritoneum.

Ascites is a collection of fluid within the peritoneal cavity, e.g. in cirrhosis of the liver or severe heart failure.

Fluid can be absorbed from the peritoneal cavity and this route can be used to introduce fluid into a dehydrated infant.

Peritoneal dialysis is a method of treating renal failure by alternately putting fluid into the cavity and letting it out again and so removing toxic substances the kidney is unable to excrete.

15

The Liver and Biliary System

The liver

The liver (Figs 15.1a, b), the largest gland in the body, weighs about 1300–1550 grams. It is reddish-brown, very vascular and soft.

It is roughly wedge-shaped with its base to the right and its apex to the left. It is situated in the upper right quadrant of the abdomen (Fig. 15.2), being protected by the ribs and costal cartilages; its lower edge reaches to the line of the costal cartilages but the edge of a healthy liver cannot be felt. It is maintained in position by the pressure of other organs within the abdomen and by 'ligaments' of peritoneum. Its smooth rounded upper surface lies beneath the diaphragm. Its visceral (postero-inferior) surface lies above the stomach, duodenum, hepatic flexure of the colon, right kidney and right adrenal gland.

Lobes
The falciform ligament of peritoneum above and an H-shaped system of fossae (with the portal fissure as the cross-piece of the H) partially divide the liver into:
large right lobe
small left lobe
quadrate and caudate lobes, smaller lobes on the visceral aspect.

Structures entering or leaving the liver at the portal fissure on the visceral surface
right and left branches of portal vein
right and left branches of hepatic artery
right and left hepatic ducts
lymph vessels
nerves

The visceral surface shows, as well as the four lobes:
inferior vena cava, which running upwards and to the right lies partly embedded in the fossa between the left lobe and the caudate lobe,
gall-bladder, which lies in the fossa between the right lobe and the quadrate lobe.

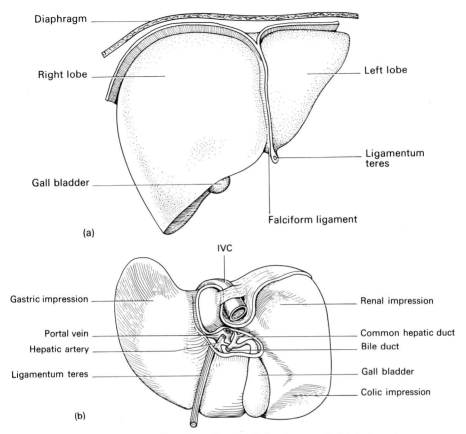

Fig. 15.1. The liver seen from (a) the front and (b) below.

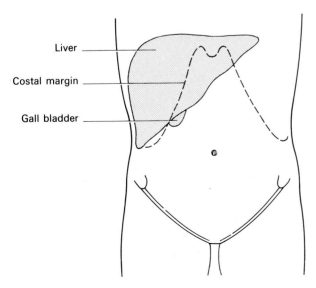

Fig. 15.2. The position of the liver.

Peritoneal covering

The peritoneum covers the whole liver except for a small area at the back of the right lobe and at the reflection of the ligaments. The *ligamentum teres* is a fibrous band connecting the visceral surface of the liver with the umbilicus; it is the remains of the left umbilical vein through which in fetal life blood passed from the placenta to the fetus.

BLOOD SUPPLY

Portal vein. About 80 per cent of the blood to the liver passes through this vein. It conveys blood from the stomach, intestine, spleen and pancreas directly to the liver. It passes upwards behind the common bile duct and the hepatic artery in the edge of the lesser omentum, a double fold of peritoneum stretching from the portal fissure to the lesser curvature of the stomach.

It divides into right and left branches which enter the liver at the portal fissure.

Hepatic artery. About 20 per cent of the blood to the liver passes through this artery. It is a branch of the coeliac artery, which is itself a branch of the upper part of the abdominal aorta. It runs in the lesser omentum to the left of the common bile duct and in front of the portal vein.

Branches:

cystic artery to the gall-bladder,

right and left hepatic arteries which enter the liver at the portal fissure.

VENOUS DRAINAGE

The venous drainage is into the two large hepatic veins, which enter the inferior vena cava at the back of the liver.

STRUCTURE

The liver is composed of:

Lobules. The liver is mainly composed of a very large number of small lobules (Fig. 15.3), each composed of hepatic cells arranged mostly in columns.

Sinusoids. These are the channels between the columns of cells and through which passes the blood from the portal vein and the hepatic artery. They are lined by endothelial cells and by cells of the reticulo-endothelial system (Fig. 15.4).

Terminal branches of the portal vein and hepatic artery on the outside of the lobules.

A *central vein* in the middle of each lobule. The veins join together to make larger veins, which unite to form the hepatic veins which open into the inferior vena cava.

Canaliculi, which run between adjacent columns of hepatic cells and unite to form the hepatic ducts.

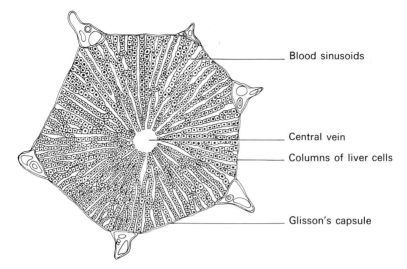

Fig. 15.3. A liver lobule.

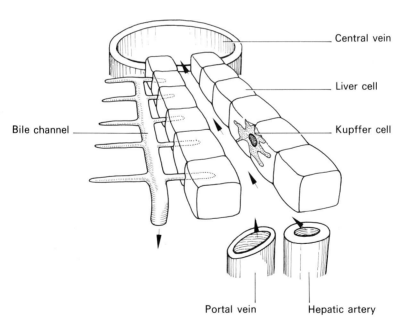

Fig. 15.4. The basic internal structure of the liver showing the direction of flow of the blood and of the bile.

Biliary system

> *Biliary system*
>   hepatic ducts, right and left
>   common hepatic duct
>   cystic duct
>   gall-bladder
>   bile duct

The *right and left hepatic ducts* emerge from the portal fissure on the visceral surface of the liver and unite to form the common hepatic duct. The *common hepatic duct* is about 3.5 cm long. The *cystic duct* runs from the end of the common hepatic duct to the gall-bladder. The *bile duct*, formed by the union of the common hepatic and cystic ducts, continues in the line of the common hepatic duct. It is about 10 cm long. It passes behind the first part of the duodenum and then either through the head of the pancreas or in a groove in it posteriorly; here it is joined by the pancreatic duct and the two open by a common opening into the second part of the duodenum; the opening is controlled by the sphincter of Oddi, a ring of smooth muscle. Sometimes the two ducts open separately into the duodenum.

The *gall-bladder* (Fig. 15.5) is a pear-shaped sac about 7.5 cm long. It can hold up to about 50 ml of bile. It consists of a fundus (the rounded end), a body and a neck. The neck is continuous with the cystic duct. The gall-bladder is adherent to the visceral surface of the liver, in the groove between the right lobe and the quadrate lobe. The fundus reaches the anterior border of the liver or projects slightly beyond it, being there in contact with the anterior abdominal wall at the tip of the ninth right costal cartilage. Its wall contains smooth muscle.

Functions of liver

The liver is the largest chemical factory in the body. It has a large blood supply ($1-1\frac{1}{2}$ litres per minute) which it receives through:
(a) the portal vein, which brings it the products of digestion from the alimentary tract,
(b) the hepatic artery, which brings it the oxygen it needs.

The blood from these two sources passes into the sinusoids between the columns of the liver cells in the lobules. Chemical exchanges take place in both directions between the blood in the sinusoids and the liver cells. Blood leaves the sinusoids by passing into the central veins and so into the hepatic veins and the inferior vena cava. Those chemical substances which are to be excreted directly from the liver pass into the canaliculi between adjacent columns of liver cells to become a constituent of bile, which is excreted through the biliary system into the intestine.

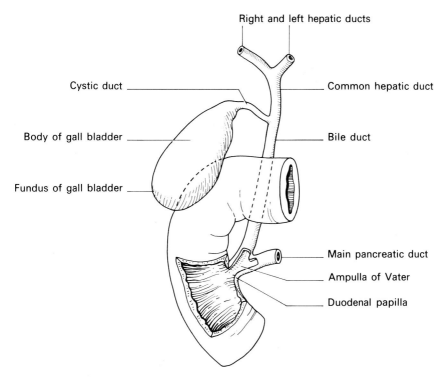

Right and left hepatic ducts

Cystic duct _____

Common hepatic duct

Body of gall bladder _____

Bile duct

Fundus of gall bladder _____

Main pancreatic duct

Ampulla of Vater

Duodenal papilla

Fig. 15.5. The gall-bladder and biliary ducts.

Functions of the liver
modification of food materials
storage of food materials
synthesis of new products
detoxication of harmful substances
formation and destruction of red blood cells

Carbohydrate metabolism
Glucose and the other monosaccharides (fructose and galactose) are converted into glycogen. Glycogen is a carbohydrate composed of many hundreds of glucose units linked together. It is a convenient way in which to store carbohydrate because:
(a) it is quickly broken down to provide energy from glucose.
(b) its energy production is high,
(c) it does not leak out of cells and does not disturb the intracellular fluid content.

Insulin, one of the two hormones of the pancreas, acts on glucose to convert it into glycogen; glucagon, the other hormone of the pancreas, converts glycogen into glucose.

Protein metabolism

Some amino acids are converted into glucose. Unwanted amino acids are converted into urea and uric acid, which are discharged from the liver cells into the blood to be excreted by the kidneys.

Lipid metabolism

When the products of lipids are required, the lipids are taken out of the fatty deposits in the body, transported in the blood to the liver, and there broken down into fatty acids and glycerol. In addition, fatty acids brought to the liver in the portal blood from the intestine are converted into types that can be used in metabolic processes.

Synthesis

The liver can synthesise:
(a) cholesterol and steroids,
(b) plasma proteins: fibrinogen, prothrombin and most of the globulins.

Storage

The liver is a store of:
(a) glycogen,
(b) fat,
(c) vitamins A, B_{12}, D, K,
(d) iron: as one of the breakdown products of haemoglobin, iron is stored in liver cells as ferritin, a protein from which iron can be released when needed.

Detoxication

The liver breaks down steroid hormones and many drugs, the breakdown products of which are excreted by the kidneys.

Formation and destruction of red blood cells

For the first 6 months of fetal life, the liver produces red blood cells; this function is subsequently performed by bone marrow. Throughout life, red blood cells are destroyed in cells of the reticulo-endothelial system, including those lining the sinusoids of the liver.

BILE

Bile is a viscous golden-yellow or greenish-yellow fluid produced continuously by liver cells in amounts of 500-1000 ml daily. It is essential for the digestion and absorption of lipids and is the medium in which are excreted certain substances which cannot be excreted readily by the kidneys.

The bile passes through the hepatic ducts and the cystic duct into the gall-bladder. As it stays there it is concentrated 5-10 times. It is discharged from the gall-bladder by the action of cholecystokinin, a hormone produced in the mucous membrane of the upper part of the small intestine whenever fat enters it.

Cholecystokinin causes contraction of the muscle of the gall-bladder and at the same time relaxation of the sphincter of Oddi, so that bile is expelled down the cystic duct and common bile duct (in which peristaltic waves occur) into the duodenum.

Constituents of bile

Bile salts. These are synthesized by liver cells from cholesterol, a steroid alcohol largely produced in the liver. The functions of the salts are to aid the digestion of fats by emulsifying them and by aiding the action of lipase, present in pancreatic juice.

Enterohepatic circulation: bile salts (and pigments) are reabsorbed from the small intestine into the portal vein and sent back to the liver to be re-used.

Bile pigments. These are mainly products of the breakdown of haemoglobin. Liver cells remove bilirubin from the plasma and excrete it in the bile. Bile pigments have no digestive functions.

Bacteria in the bowel convert bilirubin into urobilinogen. Urobilinogen is either reabsorbed from the bowel or converted into stercobilin, which is excreted in the faeces, to which it gives the brown colour.

Cholesterol.

Electrolytes: sodium, potassium, chlorides, bicarbonate.

Some drugs: salicylates, some antibiotics etc.

Clinical features

Liver cells can be damaged or destroyed and the whole range of liver functions be interfered with by many diseases:

(a) infections, especially certain viruses, e.g. the virus that causes infective hepatitis.

(b) alcohol,

(c) certain drugs,

(d) new growths.

Complete liver failure can occur. New liver cells, however, can grow to replace damaged or dead ones.

Jaundice appears when there is an excessive amount of bile pigment in the body and begins to be visible in the conjunctiva of the eyes when the plasma bilirubin exceeds 35 µmol/l. Common causes are:

(a) excessive destruction of red blood cells with an over-production of bilirubin,

(b) failure of liver function from any cause,

(c) obstruction to the flow of bile through the biliary ducts.

Stools become pale if bile pigments cannot get through to them.

Gall-stones are composed of (a) cholesterol, (b) bile pigments, (c) cholesterol and bile pigment. The reason for their formation is not clear; factors involved are concentration of bile, stasis of bile and infection of the gall-bladder. They can block the common bile duct and cause severe pain and jaundice.

16

Nutrition and Metabolism

Food provides the energy for the maintenance of vital functions, for physical activities and for growth. Protein, carbohydrate and fat are the energy. In the conversion of food into energy and heat, oxygen (O_2) is utilized and carbon dioxide (CO_2) produced. The *respiratory quotient* (RQ) is the ratio of CO_2 produced to O_2 consumed and is about 0.85.

Energy requirements

The basal metabolic rate (BMR) is the energy requirement under basic conditions—a person must be resting and not had a meal for at least 12 hours. To find the energy requirements of an individual there must be added the requirement for work and for ordinary daily activities, such as dressing, walking about, doing domestic work. Energy is measured in megajoules (MJ).

Energy expenditure of average people

Men, weight 70 kilos	Women, weight 56 kilos
sedentary work	sedentary work
10.5 MJ	8.8 MJ
light work	moderate work
12.6 MJ	10.5 MJ
heavy work	in pregnancy
18–20 MJ	10.5 MJ
	breast feeding
	12.6 MJ

Appetite

In animals and probably in man there are in the hypothalamus of the brain two centres concerned with appetite, one stimulating it, the other reducing it. Sensations of hunger appear to arise in the hypothalamus as a result of a fall in the amount of glucose in the blood passing through its capillaries.

In physically active people appetite usually controls eating in such a way that the energy intake is equal to the amount necessary to provide for the work done and to maintain weight.

Clinical features

The BMR is increased when there is over-functioning of the thyroid gland (hyperthyroidism) and decreased when there is under-functioning (cretinism, myxoedema).

Obesity is usually due to an excessive intake of calories, surplus fat and carbohydrate being stored as fat. *Under-nutrition* occurs when the daily intake provides less than 8.0 MJ a day.

FOOD

Composition of food

proteins	minerals
carbohydrates	vitamins
fats	water

Protein

Sources: meat, milk, fish, eggs, peas, beans, lentils, nuts. The average adult daily intake of protein is 40 g.

Protein is necessary for (a) growth, (b) the replacement of worn-out or damaged tissue, (c) hormones and enzymes. A protein is a complex molecule composed of simpler substances called amino acids. Some amino acids are called 'essential' because they cannot be synthesized in the body and must be in the diet. The 'non-essential' amino acids are just as necessary, but they can be synthesized in the body.

Digestion of protein

Proteins in food are acted upon by:

pepsin in gastric juice,

trypsin and chymotrypsin in pancreatic juice,

erepsin in small intestine juice.

By the actions of these enzymes proteins are broken down into the amino acids of which they are composed.

Absorption of amino acids and fate in the body

Amino acids are absorbed through the wall of the small intestine into the bloodstream. Any cell which requires an amino acid for growth or repair has the ability to abstract it out of the blood and use it to build up a new protein.

As the non-essential amino acids are interconvertible, some proteins will be the direct result of digestion, others of synthesis in the body. Protein there is in a state of being constantly broken down into amino acids and amino acids are being constantly built up into new protein. Any surplus amino acid is broken down by the liver. Ammonia is produced and converted into urea, which passes into the blood and is excreted by the kidneys.

Carbohydrate (CH)

Sources: cereal—wheat, barley, oats, maize, rice etc; sugar; potatoes. The average adult daily intake of carbohydrate is 350–450 g.

This class of food includes the sugars and starches, composed of carbon, oxygen and hydrogen. They are present in food as:

　　monosaccharides (glucose, fructose, galactose),
　　disaccharides (sucrose, maltose, lactose),
　　polysaccharides (starches).

Digestion of carbohydrates

Disaccharides and polysaccharides have to be converted into monosaccharides because only monosaccharides can be absorbed by the small intestine.

　　sucrase converts sucrose into glucose and fructose,
　　amylase converts starch into maltose,
　　maltase converts maltose into glucose,
　　lactase converts lactose into glucose and galactose.

Absorption and fate in the body

Glucose, fructose and galactose pass through the wall of the small intestine into the blood stream. Fructose and galactose are converted in the body into glucose.

Glucose is the end-product of the digestion and absorption of all carbohydrate. The blood glucose level varies with absorption, being highest after a meal. The resting level (when no food has been taken for several hours) is 4.5–8.9 mmol/l.

Glucose can pass freely in and out of the cells. It is used solely as a source of energy. It is stored as glycogen in the cells of the liver, being converted into glycogen by the action of insulin (a hormone secreted by the pancreas). Glycogen is converted back into glucose by the action of glucagon (another hormone secreted by the pancreas) and adrenaline (a hormone secreted by the adrenal glands).

Skeletal muscle contains glycogen, which is used up in the production of muscular energy. This glycogen is not convertible back into glucose.

Fat

Sources: fatty meat, nuts, butter, margarine, cream, cheese. The average adult daily intake of fat is 90–120 g.

Fats are composed of glycerol and fatty acids. The term 'lipid' is used to describe fats, sterols and some other substances.

Digestion
The digestion of fat is complicated because fats are not soluble in water. In order that they can be absorbed:

bile salts convert fat into small droplets,

lipase (an enzyme in pancreatic juice) assists the bile salts in their action and converts some of the fatty acids into monoglycerides, diglycerides, and free fatty acids.

Absorption and fate in the body
The smaller droplets of fat pass into the lymph vessels of the small intestine and reach the bloodstream through the receptaculum chyli and thoracic duct.

Some of the fat passes into the capillaries of the small intestine and thence through the portal vein into the liver.

Fat is stored or converted into energy. Any not required as energy is deposited in the fat depots of the body (mainly within the abdomen or under the skin) or is incorporated into cells. When required fat is withdrawn from the fat depots and converted into glycerol and fatty acids in the liver. The glycerol is converted into glycogen and this glycogen is used in the same way as glycogen obtained from carbohydrate. The fatty acids are oxidized into heat and energy or completely into carbon dioxide and water.

Clinical features
Ketosis is the condition in which there is a surplus of ketone bodies in the blood as a result of the imperfect oxidation of carbohydrate, fat then being used as a source of energy. Fatty acids are normally oxidized by products of carbohydrate metabolism, but these products do not become available when carbohydrate metabolism is impaired with the result that ketone bodies are produced. Ketosis is liable to occur in severe diabetes mellitus and can cause coma and death.

MINERALS AND MINERAL SALTS

Essential minerals and salts	
calcium	iron
phosphorus	magnesium
sodium	zinc
potassium	copper
iodine	cobalt

Calcium

Sources: most foods contain calcium and especially milk, butter, cheese.

Calcium is taken into the body throughout the small intestine, mostly not by absorption but by an active process in the duodenum and proximal part of the jejunum. The amount absorbed and the rate of its absorption depend upon a number of factors, including:

    bodily requirements,

    parathormone (the hormone of the parathyroid glands),

    age.

Absorption is increased during the later stages of pregnancy and during lactation, with the demands of the baby, and decreased sharply in old age.

    Calcium is present in:

(a) teeth and bone, which contain 99 per cent of the calcium in the body,

(b) blood.

The amount in the blood is controlled by:

    vitamin D, which is itself controlled by parathormone: the blood calcium is raised when it becomes too low,

    calcitonin, a hormone produced mainly or entirely by the thyroid gland: the blood calcium is lowered when it becomes too high.

    Calcium is necessary for:

    the ossification of bone and teeth,

    the coagulation of the blood,

    the contractions of the heart,

    the transmission of impulses at the neuromuscular junctions.

Phosphorus

Sources: principally milk, eggs, liver.

    Its main functions are:

(a) to provide energy for muscles in the form of ATP and creatine phosphate,

(b) to take part, as calcium phosphate, in the structure of bone,

(c) to take part in the buffer systems through which the pH of the body is controlled.

Sodium

Sources: common salt (sodium chloride); sodium citrate and sodium tartrate occur in fruit and vegetables.

    Sodium is an important constituent of cells and tissue-fluids. It is the principal extracellular cation, the amount of extracellular fluid being controlled by the amount of sodium in it.

    The body is normally in sodium equilibrium. Sodium is filtered from the plasma by the renal glomeruli, but most of it is reabsorbed in the renal tubules according to the body's needs. Salt is also excreted in sweat.

    The amount required by the body is less than 1 g daily, except when sweating is profuse, and the usual intake is much greater than the need.

Potassium

Sources: present in most foods, especially fresh orange juice.

Most of the potassium in the body is in the cells. It is the main intracellular cation and necessary for maintaining intracellular fluid volume.

Potassium is important in cell membrane activity. Its presence in nerve cells maintains nervous excitability, and the exchange of potassium and sodium ions through the membrane surrounding a nerve fibre is the basic action in the transmission of a nerve impulse. It is necessary for muscle contraction, and a shortage and an excess of potassium both produce muscle weakness, fatigue and paralysis.

Iodine

Sources: sea-food and crops grown close to the sea. Table salt can be iodized, i.e. have potassium iodide added to it; this is done in countries remote from the sea (e.g. Switzerland) where lack of iodine in the soil and therefore in the food grown on it is a cause of goitre of the thyroid gland.

Iodine is a necessary constituent of two hormones produced by the thyroid gland: thyroxine and tri-iodothyronine.

Iron

Sources: principally beef, liver, spinach.

Iron is absorbed in the duodenum and first part of the jejunum. It is absorbed mainly as simple salts, ferrous rather than ferric; small quantities of metallic iron can be absorbed. It is an essential part of the haemoglobin molecule and is therefore necessary for the formation of red blood cells.

The average diet in Britain contains 15–20 mg of iron per day. Normally about 10 per cent of this is absorbed, but up to 60 per cent can be absorbed if the body is iron-deficient.

Every day 50 ml of blood are destroyed. This amount contains 25 mg of iron, but most of this iron is used to form new red blood cells. About 1 mg of iron is lost a day in urine, in faeces (from bile) and in desquamation of the skin.

Women lose about 25–30 mg of iron in each menstrual period in addition to the other losses, and therefore require a higher intake of iron than men do. During pregnancy there is an extra demand for iron by the fetus, especially during the last three months of pregnancy.

Clinical features

An *iron-deficiency anaemia* occurs in those who from ignorance, poverty or food-faddism do not eat the main iron-containing foods. It is particularly liable to occur in women because of their extra need of iron. The red cells contain less haemoglobin than they should and are small and pale.

Magnesium

The metabolism of magnesium is closely connected with that of calcium; half the body content of magnesium is in bone. It is necessary for normal neuro-muscular activity and for the activity of smooth and cardiac muscle.

Clinical features
Lack of magnesium, which can be due to a failure of absorption in the small intestine, produces tetany (an hyperexcitability of muscle) and is a factor in the production of renal stones containing calcium.

Zinc

Zinc occurs in all living cells and in high concentrations in the liver, bone, pan-creas, prostate gland, skin and choroid coat.

It is necessary for growth, for the synthesis of proteins and nucleic acid, and for the healing of wounds and fractures.

Clinical features
Zinc-deficiency diseases can occur in malnourished people. Growth and sexual development are retarded and wound and fracture healing is prolonged.

Copper

Copper is essential for some enzyme processes.

Clinical features
In *Wilson's disease* (hepatolenticular degeneration) there is a genetically trans-mitted disorder of copper metabolism with excessive amounts of copper being deposited in the liver, brain, pancreas etc.

Cobalt

Cobalt is an integral part of the vitamin B_{12} molecule and is therefore necessary for red blood cell formation.

THE VITAMINS

The vitamins are organic compounds which in small amounts are necessary for the health of the body. They are not a source of energy.

Vitamins
   fat soluble: A, D, K
   water soluble: B complex, C

Vitamin A

Sources: fish-liver oils (such as cod-liver oil and halibut-liver oil); milk, butter, cream, eggs; liver; carrots, spinach. Margarine, made from vegetable fats, has vitamin A added during manufacture. Carotenes, which are present in green vegetables and carrots, are converted into vitamin A in the body.

Absorption. For the vitamin to be absorbed from the small intestine, the diet must contain fat. Absorption depends upon the presence of lipase, secreted by the pancreas, and bile. It is stored in the liver.

Functions. Vitamin A is necessary for:
(a) the health of epithelium; lack of it produces degeneration of mucous membranes, of the salivary, lacrimal and sweat glands, and of the conjunctiva,
(b) the regeneration of visual purple in the eye after it has been 'bleached' by light; a failure of the eyes to adapt to the dark is an early sign of vitamin A deficiency,
(c) possibly growth and development of new bone.
It may have anticancer properties.

Vitamin B complex

The original vitamin B was found to be composed of several vitamins, of which the most important are:

thiamine   niacin   riboflavin
vitamin B_{12} folic acid

THIAMINE

Sources: liver, kidney, eggs, pork, ham, yeast, beans, peas, nuts, husk of rice.

Functions. It is necessary for health. Lack of it produces beri-beri, a disease which occurs mostly in the Far East in places where the staple diet is of polished rice, i.e. rice from which the vitamin-containing husk has been removed. Clinical features of beri-beri are peripheral neuropathy, cardiac failure, mental and neurological disorders.

NIACIN

Sources: cereals, meat, liver, yeast, fish, wheat flour.

Functions. It is necessary for health. Lack of it produces pellagra, a disease characterized by dermatitis, diarrhoea, dementia. It is most likely to occur where maize is a staple article of diet for maize contains the vitamin in a form that is not liberated in the body. It can occur when there is a failure of absorption from the bowel, as can happen in the malabsorption syndromes and in chronic alcoholism.

RIBOFLAVIN

Sources: milk, cheese, eggs, green vegetables, liver, yeast. It is destroyed by sunlight and disappears from milk left standing in bottles in the sun.

Functions. It is essential for the health of epithelium. Lack of it causes degeneration of the cornea, degeneration of the mucous membrane of the mouth, and dermatitis. Deficiency is likely to be due to an inadequate diet, malabsorption from the small intestine, and the taking of a wide-spectrum antibiotic for a long time.

VITAMIN B_{12} (cyanocobalamin)

Sources: liver, kidney, muscle.

Absorption. The absorption of vitamin B_{12} from the small intestine is dependent upon the presence in the gastric juice of the *intrinsic factor*, a mucoprotein secreted by the mucous membrane of the stomach.

Function. Vitamin B_{12} is necessary for red cell formation. It contains cobalt. In pernicious anaemia the mucous membrane of the stomach is damaged by antibodies, intrinsic factor is not produced, and vitamin B_{12} is not absorbed. Total gastrectomy reduces the amount of intrinsic factor. Strict vegetarians (vegans) may develop vitamin B_{12} deficiency because there is little of it in fruit and vegetables.

FOLIC ACID

Sources: liver, green vegetables, yeast.

Function. It is necessary for red cell formation.

Vitamin C (ascorbic acid)

Sources: fresh fruit and vegetables, especially oranges, tomatoes, grapefruit, blackcurrants, watercress, rose-hips. There is not much in apples, pears, plums. Vitamin C is destroyed by cooking, especially in alkaline solutions, e.g. when soda has been added to the cooking water, and by prolonged heating. Pasteurization destroys vitamin C in milk.

Functions. Large amounts of it are contained in the adrenal cortex, but elsewhere there is little stored in the body. It is necessary for the health of capillary walls, for the cement substance which binds cells together, and for the prompt healing of wounds. Lack of it causes scurvy, which is characterized by haemorrhages and anaemia. Scurvy is likely to occur in babies fed on pasteurized milk only, old people living alone and not eating enough fruit and vegetables, and people living in institutions where cooked food is left for long periods in heated containers.

Vitamin D

Sources: liver, butter, vitaminized margarine, cheese, eggs, fish-liver oils, the flesh of fatty fish (herring, salmon etc, identifiable by their pinky-brown colour).

Active vitamin D is found in few natural substances, but precursors of it called provitamins (belonging to the chemical class of sterols) are found in plant and animal tissues.

Plant provitamin D (ergosterol) acquires vitamin D activity when exposed to ultraviolet light and is then called vitamin D_2 (calciferol). *Animal provitamin D* is present in the skin and acquires vitamin D activity when exposed to ultra-violet light and is then called vitamin D_3 (cholecalciferol).

The vitamin D in food is mostly vitamin D_2, which after absorption from the small intestine is supplemented by smaller amounts of vitamin D_3 from the skin. It is thought that neither D_2 nor D_3 is biologically active and that they have to be converted in the liver into a related substance known as 25-HCC. This is then converted in the kidney into another form known as 1,25-DHCC, which is the biologically active form of vitamin D.

The production of 1,25-DHCC is probably controlled by the parathyroid glands. It acts like a hormone and appears to be responsible for:
(a) the absorption of calcium by the small intestine,
(b) the mobilization of calcium from bone into the blood.

Lack of vitamin D produces *rickets* in children and *osteomalacia* in adults. Both are disorders of calcium metabolism and characterized by a softening of bones and the production of deformities.

Vitamin K

Sources: meat, green vegetables (Brussels sprouts, spinach). It is synthesized by bacteria in the intestines.

Function. It is essential for the formation of prothrombin by the liver and therefore for the clotting of blood.

17

The Skin

The skin is one of the largest organs of the body: it forms 15 per cent of the total body-weight.

```
Skin
   epidermis
   dermis
```

Epidermis

The epidermis, the outer layer, is composed mainly of stratified squamous epithelium. The cells composing it are continually being formed from a deep germinal layer of columnar epithelium and become flattened as they are pushed by new cells towards the surface where they are worn off by friction. The outer layers contain keratin, a horny protein; there is little of it on the surfaces slightly subject to wear and tear, such as the inner surfaces of the arm and thigh, and much more on the extensor surfaces; it is especially thick on the soles of the feet.

Pigmentation of the skin is mainly due to the presence of melanin, a black pigment, in the deeper layers of the epidermis. The more melanin the darker the skin. Pigmentation is mainly controlled by adrenal and pituitary hormones. Pigmentation is increased by ultraviolet light.

Dermis

The dermis is a layer composed of collagen and fibrous and elastic tissue. The superficial layer projects into the epidermis in a number of small papillae. The deeper layer rests on subcutaneous tissue and fascia. It contains blood vessels, lymph vessels and nerves.

BLOOD VESSELS

Plexuses of capillaries are present beneath the epidermis, beneath the dermis, in the papillae, and around the sweat and sebaceous glands and hair follicles. There are thoroughfare vessels present in some parts of the dermis; in them

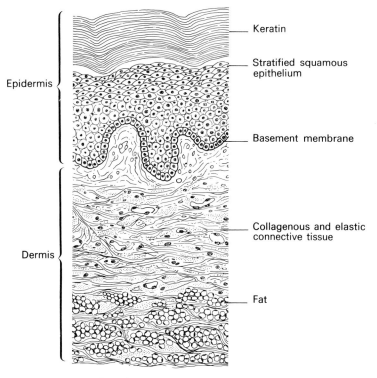

Fig. 17.1. Section through skin.

blood can pass directly from arteriole to vein without having to go through any capillaries.

NERVES

The skin is supplied with sensory and sympathetic nerves.

Sensory nerve fibres end in the skin in various forms:
(a) *Free nerve-endings,*
(b) *A plexus of nerves* around the hair follicles,
(c) *Meissnerian corpuscles* (Fig. 17.2a), small encapsulated structures around nerve-endings in the papillae,
(d) *Paccinian corpuscles* (Fig. 17.2b), larger encapsulated structures found deep to the dermis.

Sympathetic fibres supply the arterioles, sweat glands and arrector pili muscles.

Nails

Nails are specialized pieces of keratinized epidermis. The nail matrix is an area of germinal cells from which grows the root of the nail. The nail bed below the

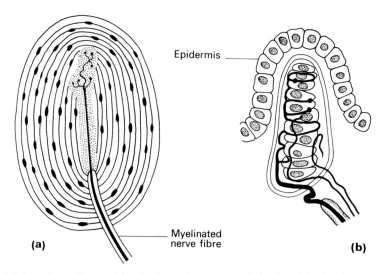

Fig. 17.2a. A touch-sensitive Meissnerian corpuscle in the skin. **b.** A pressure-sensitive Paccinian corpuscle in the skin.

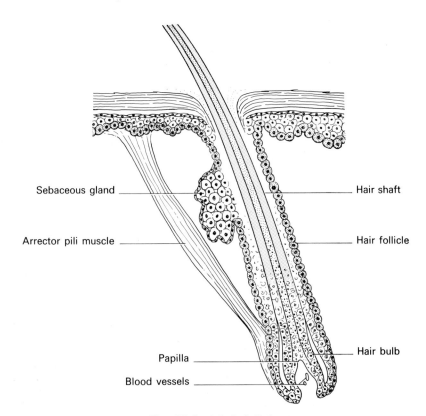

Fig. 17.3. A hair follicle.

nail is a thin layer of epidermis. The pulp of the finger below the nail bed is formed of loose vascular connective tissue.

Hair

A hair is a keratinized outgrowth from a papilla at the bottom of a hair follicle (Fig. 17.3), which is a narrow tube running from the surface of the skin through the epidermis to the dermis. The colour of a hair is determined by the amount of melanin in its outer layers: the more melanin the darker the hair.

An *arrector pili muscle* is a thin bundle of smooth muscle fibres attached at one end to a hair follicle and at the other to connective tissue in the dermis. It is supplied by sympathetic nerves.

Sebaceous glands

Sebaceous glands occur in all the skin except that of the palms and soles.

A sebaceous gland is a small gland wedged between a hair follicle and its arrector pili muscle. It opens by a duct into the upper third of the follicle.

Sebum, the product of a sebaceous gland, is the result of fatty degeneration of its cells. It is squeezed out of the gland and into the follicle at every contraction of the arrector pili muscle. It protects and lubricates the skin and hair.

Sweat glands

A sweat gland is a single coiled tube situated in subcutaneous tissue and with a long duct which opens on the surface of the skin. There are two kinds of sweat glands:

the *ordinary sweat glands*, innervated by sympathetic nerves,

apocrine glands in the axilla, vulva and nipples. They have no nerve supply; and stimulated by adrenaline they produce a yellowish secretion with a characteristic smell (partly due to bacterial action on it). These glands are inactive before puberty and their smell may be a sexual attraction.

The *ceruminous glands*, the wax-producing glands in the external auditory meatus, are modified sweat glands.

Creases and ridges. The creases of the palms and soles are formed about the 12th week of embryonic life. The main creases tend to develop where the palms and soles will later become flexed. They vary in position in different people, and they show differences in length and depth at different periods of life in the same person. The *ridges* are the basis of identification by finger, palm or sole prints. They are narrow ridges arranged in groups of parallel lines in loops and other patterns. They appear about the same time as the creases, but they are constant throughout the life of a person and are not altered by superficial damage to the skin.

FUNCTIONS OF THE SKIN

Functions of the skin
  protection
  sensation
  heat regulation
  storage
  absorption

Protection

The skin protects the internal structures of the body against trauma and against invasion by harmful micro-organisms. Most micro-organisms have difficulty in penetrating intact skin but can get in through cuts and abrasions. In addition to the protection given by the horny layers, an additional protection is provided by the acidity of sweat and the presence of fatty acids in sebum, which inhibit the growth of micro-organisms, and by the action on the harmful micro-organisms of the harmless micro-organisms normally present on the surface of the skin.

Sensation

Sensations of touch, pain, changes of temperature and pressure are appreciated in the skin and subcutaneous tissues, and transmitted through sensory nerves to the spinal cord and brain.

The precise functions of the various nerve-endings in the skin are not definitely known. One of the problems is that in any area of skin at any one time there are separate tiny areas in which touch, pain and changes of temperature are individually appreciated, but these areas change from day to day, an area which reacted to heat one day reacting to touch another, and so on. Possibly not all the nerve-endings are transmitting at any one time; some may be resting.

The deeper lying paccinian corpuscles are thought to react to deep pressure.

Heat regulation

See Chapter 18.

Storage

The skin acts as a store of water and fat, which can be drawn upon in need.

Absorption

The skin can absorb:
(a) ultraviolet light, which acts upon a vitamin D precursor,
(b) certain drugs applied as ointments etc.

18

The Regulation of Body Temperature

For temperature purposes the body can be considered as:
(a) a peripheral shell (skin, subcutaneous tissue, muscle) and limbs,
(b) an inner core (contents of chest, abdomen, skull).

The temperature of the peripheral shell can vary, but the temperature of the inner core has to be kept fairly constant. Man maintains this equilibrium in spite of wide variation in the temperature of his environment (from the arctic to the tropics) by stabilizing heat gain and heat loss.

Heat gain

Heat is gained by being:
    produced in the body,
    taken up from the environment.

HEAT PRODUCTION

Heat is produced by all the metabolic activities of the body. The amount produced by internal organs (the liver, heart etc) is fairly constant; the amount produced by the skeletal muscles varies from a little at rest to a great deal during exercise. Shivering is another way of producing heat.

There is a relation between heat production and the surface area of the body. When the surface area of the body is relatively large (as in infants), there is a need for relatively greater heat production to compensate for the greater heat loss.

HEAT FROM THE ENVIRONMENT

The body takes in heat from things hotter than itself:
    direct radiation from the sun,
    reflected radiation from the sky,
    hot food, hot drinks, hot baths,
    hot air in hot climates,
    hot soil in contact with the body.

Heat loss

Heat is lost in three ways.

FROM THE SKIN

Heat is lost from the skin by conduction, radiation and convection, by invisible perspiration and the evaporation of sweat. This loss is controlled by variations in the amount of blood passing through the skin, produced by alterations in the size of the blood vessels in it. Heat loss through the skin is affected by the amount and kinds of clothes worn.

Conduction is a direct loss from one object to a colder object. *Radiation* is a spread of heat from the skin to the colder air. *Convection* varies with the flow of air across the skin, e.g. when moved by the wind or a fan.

IN EXPIRED AIR

It is saturated with water vapour at body temperature.

IN THE URINE AND FAECES

Control of temperature by water evaporation

The most important loss of heat is by the evaporation of water from the surface of the skin. Water and heat are lost from the skin in two ways:

INSENSIBLE PERSPIRATION

About 240 ml of water diffuse through the skin every 24 hours. It is called 'insensible' because it can neither be felt nor seen. Its diffusion is continuous and not much affected by the environment. About 140 kcal of heat are lost in this way in 24 hours.

SWEAT

Sweat contains sodium chloride, urea and lactic acid in dilute solution. It is secreted from the ordinary sweat glands, which are distributed all over the skin. It is secreted as a result of dilatation of skin vessels under the nervous control of the hypothalamus, the cerebral cortex and other parts of the central nervous system. Sweating is increased by:
(a) a rise in temperature of the body,
(b) emotional states,
(c) exercise,
(d) fainting, nausea, vomiting, a low blood sugar.
    In extreme conditions of heat up to 1.7 litres of sweat (equivalent to

1000 kcal) can be lost in 1 hour, or up to 12 litres of sweat (equivalent to 7000 kcal) in 24 hours.

Heat is lost only if the sweat is allowed to evaporate. If it is wiped away, there is no heat loss, only a loss of fluid from the body.

The normal range of temperature

The temperature is maintained between 36–37.5°C (97–99.5°F). There is in most people a daily alteration between a low morning temperature and a high evening temperature, with the minimum temperature in the small hours and the maximum in the afternoon. The pattern is characteristic for an individual and does not vary with the seasons. It does not alter if he or she works at night. In women there is a monthly variation, the temperature during the first half of the menstrual cycle being lower than that during the second half. There is a sudden rise of about 0.5°C at the time of ovulation.

The regulation of temperature

Temperature is regulated by the nervous system and by the endocrine system.

NERVOUS SYSTEM

(a) Cooling and heating the skin stimulates temperature-sensitive nerve-endings with the production of appropriate responses—shivering for cold, sweating for heat.

(b) The hypothalamus in the brain responds to the temperature of the blood passing through its capillaries. It contains two centres for heat regulation. One responds to a rise in temperature by causing vasodilatation and so heat loss. The other responds to a fall in temperature by causing vasoconstriction and an activation of further heat production. Through its connections within the brain, the hypothalamus receives stimuli from the thalamus and can through the autonomic nervous system modify pulmonary activity (seen best in a dog's panting when hot), sweat secretion, glandular and muscular activity.

ENDOCRINE SYSTEM

Adrenal medulla: cold increases the secretion of adrenaline, which stimulates metabolism and so increases heat production.

Thyroid gland: cold increases the secretion of thyroxine, with an increase in metabolism and heat production.

Exposure to heat

Exposure to heat produces:
   increased blood flow through the skin,
   increased sweat production.

Exposure to cold

Exposure to cold produces:
    shivering,
    vasoconstriction as a result of cooler blood passing through the hypothalamus,
    less blood passes through the skin, less heat is lost, less sweat is produced,
    increased adrenaline and thyroxine secretion.

Clinical features
Fever is due to:
(a) infections,
(b) destruction of tissues in certain diseases.

It is the result of a failure of normal temperature-regulating mechanisms, possibly as the result of the action on them of an antigen from leucocytes.

Fever usually passes through three stages.
(i) An attack of shivering; severe shivering is called a rigor. Skin vessels are constricted and loss of heat is reduced to a minimum.
(ii) The temperature rises; skin vessels are dilated, the sweat glands remaining usually inactive; metabolic processes are speeded up and there is a greater production of heat.
(iii) The temperature falls, heat loss being greater than heat production; profuse sweating occurs.

Hyperthermia (excessive heat) occurs when the temperature of the air is greater than the temperature of the skin and when there is a total failure of heat-controlling mechanisms. *Heat stroke* occurs in men working under very hot conditions; hyperpyrexia, dehydration and loss of salt occur. *Hypothermia* is excessive reduction of temperature. It can occur when a person is exposed to extremes of cold, is inadequately clad or has a deficiency of thyroid hormone secretion. It can be a cause of death in old people living alone with inadequate heat and food. Hypothermia is deliberately produced for certain operations by cooling the blood or skin in order to reduce metabolic processes to a minimum.

19

The Lymphatic System

Lymphatic system
  lymph capillaries
  lymph vessels
  lymph nodes
  other lymphatic areas
  lymph

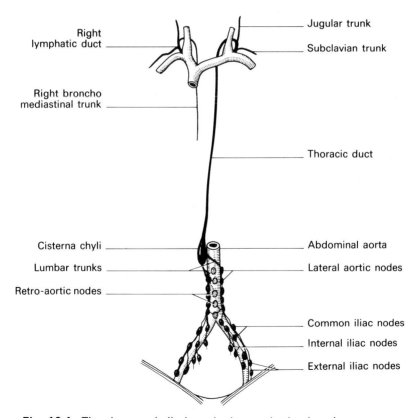

Fig. 19.1. The cisterna chyli, thoracic duct and other lymph structures.

LYMPH CAPILLARIES

A series of thin-walled blind-ended tubes lying in the tissue-spaces of various organs and tissues.

LYMPH VESSELS

These are larger tubes formed by the union of lymph capillaries; they are transparent and their many valves give them a beaded appearance. Superficial lymph vessels drain the skin; deeper lymph vessels drain the other structures of the body. Lymph vessels pass towards and enter lymph nodes.

There are no lymph vessels in:
central nervous system,
striated muscle,
non-vascular structures—nails, hair, cornea, cartilage.

LYMPH NODES

Round or oval masses of lymph cells, surrounded by a capsule (Fig. 19.2). They are soft and greyish pink; those draining the lungs of town-dwellers become black with a deposit in them of carbon which has been breathed in. Lymph vessels enter and leave them, connecting one node with another. They are often called 'lymph glands', but this is an incorrect term as they do not secrete anything.

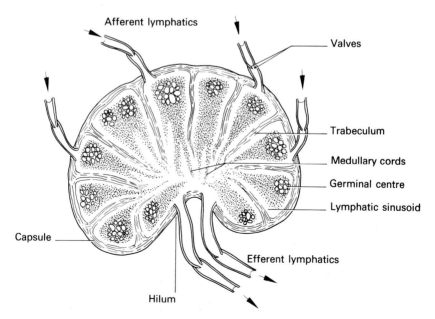

Fig. 19.2. Section through a lymph node.

OTHER STRUCTURES

Other structures in the lymphatic system are composed wholly or in part of lymph tissue identical with that in the lymph nodes:

Peyer's patches in the small intestine,
tonsils,
nasopharyngeal tonsil,
thymus gland.

The total weight of lymph tissue in the body is about equal in weight to that of the liver.

LYMPH

The fluid in the lymph vessels. It is similar in chemical composition to blood plasma and contains large numbers of lymphocytes passing along the vessels

Lymph drainage of organs and tissues	
Organs and tissues	*Site of drainage*
head and neck	nodes in neck
arms	nodes in axilla, then nodes in neck
breast	nodes in axilla, then nodes in neck
	nodes within chest along internal mammary artery
	lymph vessels in rectus abdominis sheath
lungs	nodes in hilum of lung, at bifurcation of trachea, around trachea
stomach	nodes around lower end of oesophagus, along greater and lesser curvatures
small intestine	mesenteric nodes, nodes in front of aorta
large intestine	nodes along mesenteric arteries and in front of aorta
pelvic contents	nodes close to lateral wall of pelvis and in front of lower end of aorta
testis and epididymis	nodes in front of aorta (near origin of testicular arteries)
external genitalia	nodes in groin
legs	nodes in popliteal fossa and groin, and in front of aorta

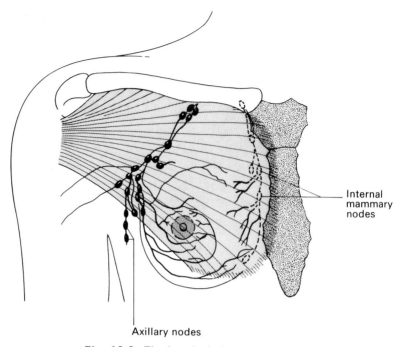

Fig. 19.3. The lymph drainage of the breast.

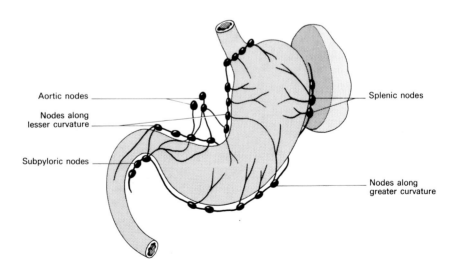

Fig. 19.4. The lymph drainage of the stomach.

from the lymph nodes in order to enter the bloodstream. Lymph vessels draining the small intestine are called *lacteals* because when fat is absorbed from the intestine much of it passes along them and gives them a milky appearance. Lymph moves along partly by being pulled by the negative pressure inside the chest, partly by being pumped along by the contraction of muscles.

Cisterna chyli and thoracic duct

The cisterna chyli (Fig. 19.1) is a large sac at the back of the upper part of the abdomen, lying behind the aorta and in front of the bodies of the upper two lumbar vertebrae. Lymph from the abdomen and legs passes through the aortic glands and thence into the cisterna chyli.

The thoracic duct (Fig. 19.1) is a large lymph vessel which is continuous with the cisterna chyli, passes through the diaphragm, runs up the back of the thorax, inclines to the left, and in the neck arches over the apex of the left lung to end by opening into the junction of the left subclavian and left internal jugular veins. It receives lymph vessels from the left arm, the left side of the head and neck, and the chest.

On the right side of the neck a short lymph vessel called the right lymph duct receives lymph vessels from the right arm and right side of the head and neck, and opens into the junction of the corresponding veins on the right side.

Functions of lymphatic system

Functions of lymphatic system
drainage of tissue spaces
absorption of fat
production of lymphocytes
production of antibodies

Drainage of the tissue spaces
Fluid and chemical substances which have passed out of the capillaries into the tissue-space pass into the lymph capillaries. Any excess of interstitial fluid is removed in lymph. Some of this fluid is reabsorbed into the blood as it passes through the capillaries in the lymph nodes. Carbon and other particles too large to be absorbed into the blood are absorbed into the lymph stream and pass along with the fluid into the lymph nodes.

Absorption of fat
About two-thirds of the fat absorbed from the small intestine pass into the lymph vessels of the small intestine, which because of their milky appearance when they are engaged in this function are called lacteals, and eventually into the blood via the cisterna chyli and thoracic duct.

Production of lymphocytes
Lymphocytes are produced in the lymph nodes and pass along the lymph vessels into the thoracic duct into the blood.

Antibody production
The lymph nodes are probably sites of antibody production.

Clinical features
The lymphatic system is one of the defences of the body against infection. Some infections are particularly liable to travel along lymph vessels into lymph nodes, which become enlarged and inflamed. Antibodies are probably manufactured in lymph nodes. Cancer cells are liable to travel along lymph vessels draining the part the cancer has arisen in, and to invade lymph nodes, which become large and hard; surgeons and radiotherapists have to include these routes in plans of treatment by surgery or radiotherapy.

20

The Endocrine Glands

The endocrine glands are glands that secrete chemical substances called hormones directly into the bloodstream. Some organs have a dual function: they produce hormones from one lot of cells and other substances from another (e.g. the pancreas produces insulin and glucagon, two hormones, and also pancreatic juice).

Endocrine glands

hypothalamus–pituitary gland complex	ovaries
thyroid gland	placenta
parathyroid gland	testes
adrenal glands	kidneys
pineal gland	pancreas
stomach	small intestine

HYPOTHALAMUS–PITUITARY GLAND COMPLEX

The hypothalamus and the pituitary gland (Fig. 20.1) are closely linked anatomically and functionally.

The *hypothalamus* is a small area of brain below the third ventricle and behind the optic chiasma; it extends downwards into the pituitary stalk. It consists of nerve cells grouped into nuclei, with nerve fibres running to it from other parts of the brain and nerve fibres running from it down the stalk of the pituitary gland into the posterior lobe of the gland. Capillaries run directly downwards from the hypothalamus to the anterior lobe of the pituitary gland.

The *pituitary gland* lies in the pituitary fossa (sella turcica), a deep depression in the upper surface of the body of the sphenoid bone. A sheet of dura mater covers the opening of the fossa. The stalk of the pituitary gland connects the hypothalamus with the gland, passing through a hole in the sheet of dura mater.

The optic chiasma of the 2nd cranial (optic) nerve lies in front of and above the pituitary gland.

The gland consists of two lobes, anterior and posterior, with different origins, structure and functions.

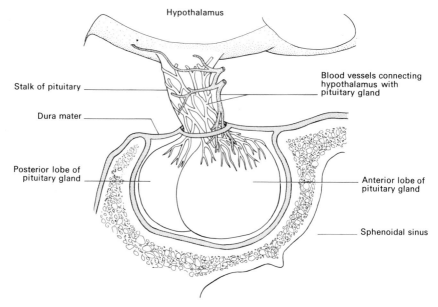

Fig. 20.1. The hypothalamic–pituitary complex.

The *anterior lobe* consists of columns of cells, which branch irregularly and are separated by sinusoids through which the blood circulates. Three types of cells can be distinguished by staining methods:

acidophils which stain red,
basophils which stain blue,
chromophobes which stain poorly.

The *posterior lobe* is smaller than the anterior lobe and is composed of nerve fibres, neuroglia and blood vessels. Nerve fibres run to it from the hypothalamus.

Hormones of anterior lobe
growth hormone
thyroid-stimulating hormone
adrenocorticotrophic hormone
gonadotrophic hormones
prolactin

Hormones of posterior lobe
antidiuretic hormone
oxytocin

Anterior lobe

The anterior lobe has been called the 'leader' or 'master-gland' of the endocrine system because of the effects its hormones have on other endocrine glands. But its activities are controlled by chemical factors, secreted in the hypothalamus and passed down into the pituitary gland mainly in the capillaries of the stalk of the gland.

Growth hormone (GH): causes nitrogen to be retained in the body and is essential for growth. It is secreted in adults as well as in children and adolescents, and has effects on fat and carbohydrate metabolism and anti-insulin properties.

Thyroid-stimulating hormone (TSH): stimulates the thyroid gland to produce thyroxine and tri-iodothyronine. Its production by the hypothalamus and its liberation from the gland is controlled by the concentration of thyroxine in the blood.

Adrenocorticotrophic hormone (ACTH): stimulates the cortex of the adrenal glands to produce glucocorticoids. Its secretion is controlled by the amount of cortisol in the blood.

Gonadotrophic hormones (gonadotrophins): act on the sex glands. In *men*, the *interstitial cell-stimulating hormone* (ICSH) stimulates the interstitial cells of the testis to produce androgens. In *women*, (a) a *follicle-stimulating hormone* (FSH) causes ripening of the ovarian follicles in which ova develop, and (b) a *luteinizing hormone* (LH), which is the same as ICSH, combines with FSH to complete the ripening of the follicles and stimulates the development of the corpus luteum. At a critical point, as FSH diminishes and LH increases, ovulation occurs. The ripening follicle secretes oestrogens, and after ovulation the corpus luteum secretes oestrogens and progesterone. The oestrogens, by a feedback mechanism, inhibit the production of FSH.

Prolactin: this hormone, produced in the pituitary gland and not in the hypothalamus, is involved in the stimulation and maintenance of lactation in the breast.

Clinical features
Enlargement of the gland by a tumour causes:
1. A ballooning of the fossa, visible on X-ray,
2. decalcification of the clinoid processes of the sphenoid bone, which project upwards just in front of and behind the fossa,
3. pressure on the optic chiasma, causing bitemporal hemianopia (blindness for objects at the side of a person) due to involvement of the nerve fibres that cross from one side to the other in the chiasma.

Under-production of the growth hormone is a cause of dwarfism.
Over-production of the growth hormone causes:
gigantism—excessive height and long arms and legs if over-production occurs in childhood or adolescence, before the epiphyses fuse with the shafts of bone and so while growth is still possible

acromegaly, after growth has finished, a condition in which jaws, hands and feet enlarge, skin is thickened.

Cushing's syndrome is due to an over-secretion of hormone by the cortex of the adrenal gland, often as a result of over-secretion of ACTH, the trunk and face become fat, hair grows excessively, blood pressure is raised. *Hypothyroid-ism* can be due to a deficiency of TSH.

Posterior lobe

The hormones are produced in the hypothalamus and pass down the nerve fibres to the posterior lobe of the pituitary gland.

Antidiuretic hormone: stimulates the distal tubules of the kidney to reabsorb water from the fluid in them.

Oxytocin: is involved in uterine action at birth (its function is not clear) and contractions of the muscle of the ducts of the breast, causing the milk to be squeezed from the deep to the superficial ducts.

Clinical feature

Diabetes insipidus is due to a failure of secretion of the antidiuretic hormone: water is not reabsorbed by the distal renal tubules and the patient passes large amounts of urine and becomes dehydrated.

THYROID GLAND

The thyroid gland (Fig. 20.2) is situated in the neck and consists of right and left lobes connected by a narrow isthmus. The two lobes are moulded round the sides of the upper part of the trachea and oesophagus; the isthmus is a narrow strip of thyroid tissue connecting the lobes across the front of the 2nd and 3rd

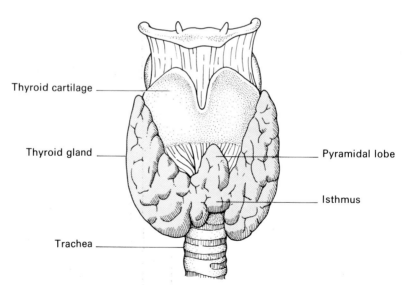

Fig. 20.2. The thyroid gland. A pyramidal lobe (of varying size) may or may not be present.

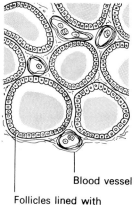

Follicles lined with
cuboidal epithelium
containing colloid

Blood vessel

Fig. 20.3. Follicles and
blood vessels of the
thyroid gland.

tracheal cartilages. The upper poles of the lobes reach as high as the thyroid cartilage. The gland is soft and brown and enclosed within a capsule. It is composed of a large number of follicles (Fig. 20.3), which consist of a single layer of cells enclosing colloid, a yellow fluid which is the major site of concentration of iodine in the gland.

The thyroid gland secretes (a) thyroxine and tri-iodothyronine and (b) thyrocalcitonin.

Thyroxine and *tri-iodothyronine*, both of which contain iodine, are essential for oxidative processes in metabolism. Their production is stimulated by the thyroid-stimulating hormone (TSH) of the anterior lobe of the pituitary gland whenever the blood thyroxine level falls.

Thyrocalcitonin is a hormone secreted by certain cells called C cells. It reduces the amount of calcium in the blood plasma. Its production is probably stimulated by a rise in the amount of calcium in the plasma. Its action is the opposite of that of parathyroid hormone, produced by the parathyroid glands.

Thyroid cells take out of the blood passing through the gland all the iodine they need. The take-up can be measured using radio-active iodine (^{131}I), the radiation from which can be measured outside the body.

Clinical features
Hypothyroid states occur when there is an inadequate secretion of thyroxine. *Cretinism* is hypothyroidism in infancy; it is due to a shortage of iodine in the diet or to a genetically determined enzyme defect; features are a failure of physical and mental growth, dwarfism, low temperature, slow pulse, thickened skin. *Myxoedema* is a hypothyroid state developing in adult life; features are diminished metabolism, obesity, slow pulse, low temperature, loss of hair, and mental deterioration. *Hyperthyroidism* occurs when there is an excessive secretion of thyroxine; it produces increased metabolism, loss of weight, irregular

activity of the heart, emotional instability. A *goitre* is an enlargement of the thyroid gland and can be due to iodine deficiency, a genetic defect, treatment by various drugs (e.g. PAS used in the treatment of tuberculosis), and in hyperthyroidism.

PARATHYROID GLANDS

The four little parathyroid glands (Fig. 20.4), each about 3 mm in diameter, lie behind the thyroid gland or are embedded in its capsule, as an upper pair and a lower pair. They can vary in size and number, and are sometimes found inside the thyroid gland or behind the pharynx or in the thorax. They are composed of clumps of cells, separated by connective tissue and with sinusoids for blood running around the cells.

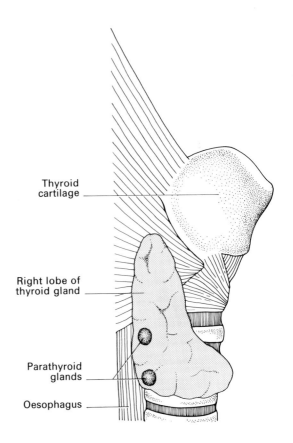

Thyroid cartilage

Right lobe of thyroid gland

Parathyroid glands

Oesophagus

Fig. 20.4. Two parathyroid glands within a lobe of the thyroid gland.

Parathyroid hormone raises the amount of calcium in the blood plasma by (a) transferring calcium from bone into plasma, (b) increasing the reabsorption of calcium by the tubules of the kidney, so that less is excreted in the urine, (c) promoting the absorption of calcium by the intestine. An increase of plasma calcium reduces the secretion of parathyroid hormone and increases that of thyrocalcitonin secreted by the thyroid gland.

Clinical features

Hyperparathyroidism is the condition in which there is an excessive secretion of the hormone, usually as a result of the growth of a tumour of parathyroid tissue; an excessive amount of calcium is transferred from bone into blood, more calcium is excreted in the urine, renal stones may develop. *Hypoparathyroidism* is the result of too little secretion of the hormone; the amount of calcium in the blood is reduced, with the production of *tetany*, in which occur muscular twitching, and painful sensations in the limbs. In operations to remove the thyroid gland a surgeon leaves its posterior part so that he does not remove the parathyroids.

ADRENAL GLANDS

The adrenal (suprarenal) glands lie at the back of the abdomen and immediately above the kidneys (Fig. 20.5), to the upper pole of which each is moulded. They are about 5 cm high, 2.5 cm wide at the base, and 1 cm thick; the left is flatter than the right and more half-moon shaped. Each is composed of a yellow cortex and an inner reddish-grey medulla. The cortex and medulla have different origins, structure and functions.

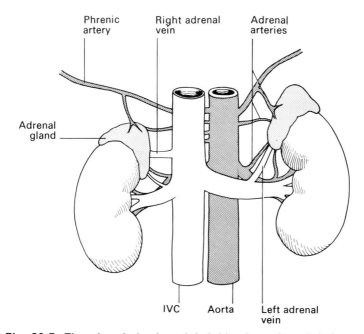

Fig. 20.5. The adrenal glands and their blood supply and drainage.

The *cortex* shows three zones of cells, an outer one of cells in clusters, a middle one of cells in columns, an inner one of cells in irregular columns. The cells have a high concentration of cholesterol which is necessary for the formation of adrenal steroids.

The *medulla* is developed from the same tissue as nervous tissue and is really a part of the sympathetic nervous system. It is composed of irregular cords of cells, surrounded by blood sinuses and innervated by sympathetic nerves.

Adrenal cortex

The cortex produces three groups of hormones with a common basic structure.

GLUCOCORTICOIDS

The secretion of glucocorticoids is regulated by ACTH from the pituitary gland. Cortisol (hydrocortisone), the most important of them, (i) is an antagonist of insulin, causes glycogen to be deposited in the liver, raises the blood sugar, and inhibits glucose-uptake by the tissues, (ii) breaks down tissue-proteins, which are converted in the liver into glycogen, (iii) is involved in the control of the exchange of water and electrolytes between cells and extracellular spaces.

MINERALOCORTICOIDS

Aldosterone regulates the balance of sodium in the body by acting on the tubules of the kidneys. It promotes the retention of sodium and the excretion of potassium. Its secretion is regulated by the level of plasma potassium and the production of renin by the kidney.

Corticosteroid is a term used to describe both glucocorticoids and mineralocorticoids.

ANDROGENS

These are produced in males, being responsible for the development of the secondary male sexual characteristics (growth of hair on the face, deepening of the voice). Their actions are weaker than those of testosterone.

Clinical features
Addison's disease, in which the cortex atrophies or is destroyed by disease, is the result of a deficiency of corticosteroids: features are loss of appetite, nausea, vomiting, a low blood sugar, pigmentation of the skin and mucous membranes, a loss of sodium from the body, a low BP and dehydration. *Cushing's syndrome*: see pp. 211–12. *Deficiency of aldosterone* causes excessive loss of water in the urine, dehydration and a low BP. *Excessive production of androgens* produces excessive sexual precocity in boys and the development of male characteristics in girls and women.

Adrenal medulla

The medulla produces adrenaline and noradrenaline. They are slightly different chemically and have similar but not identical actions. Their secretion, produced as a response to stress, acts as a boost to the sympathetic system and enables the body to take effective action in a dangerous or potentially dangerous situation.

Actions of adrenaline and noradrenaline		
organ	*adrenaline*	*noradrenaline*
heart	rate, force & output increased	rate increased, then decreased; little action on force & output
coronary areteries	dilated	constricted
blood vessels in voluntary muscle	dilated	constricted
blood vessels in skin and viscera	constricted	constricted
blood pressure	rise, then fall due to dilatation of blood vessels in muscle	rise
involuntary muscle	tone & peristalsis in gut decreased; sphincters contracted; bronchi dilated	tone & peristalsis in gut decreased; sphincters contracted
metabolism	oxygen consumption increased; glycogen converted to glucose; rise of blood sugar	little action

Clinical feature
Tumours of the medulla are rare; they cause a high BP, sweating, glycosuria (sugar in the urine) and anxiety.

TESTIS

In addition to producing spermatozoa the testis produces androgens, the male sex hormones. It does this in the interstitial cells, certain large cells in the tissue between the tubules.

Testosterone is the most important of the androgens. It is produced in the following way: the hypothalamus of the brain can appreciate the level of androgens in the blood, it responds to a fall in the level by stimulating the anterior lobe of the pituitary gland to produce its interstitial cell-stimulating hormone (ICSH), this hormone circulates in the blood and stimulates the interstitial cells of the testis to produce androgens.

Testosterone causes development of secondary male sexual characteristics and builds up proteins from amino acids.

Clinical features
A failure of testicular function can be due to (a) a failure of ICSH production in the pituitary gland, (b) failure of the testis due to genetic abnormalities, injury or disease (e.g. mumps). Excessive production of androgens causes precocious puberty.

OVARY

In addition to its function as the producer of ova, the ovary produces endocrine hormones from the cells of the ovarian follicle and its successor the corpus luteum (see p. 282).

Oestrogens are hormones produced in the ripening ovarian follicle and the corpus luteum. They are responsible for the development of the female secondary sexual characteristics and for producing cyclical changes in the endometrium of the uterus, in the cervix of the uterus and in the vagina.

The follicle-stimulating hormone (FSH) of the anterior lobe of the pituitary gland stimulates the production of oestradiol from progesterone.

Progesterone is a hormone secreted by the corpus luteum in response to the secretion of the luteinizing hormone (LH) from the anterior lobe of the pituitary gland. The corpus luteum is an ovarian follicle after the ovum has been discharged from it; and if a woman becomes pregnant it increases in size up to the 3rd month of pregnancy after which its functions are taken over by the placenta. The secretion of progesterone from it is greatest therefore during the first 3 months of pregnancy. Progesterone stimulates endometrial development, inhibits uterine movements, enables pregnancy to continue by inhibiting menstrual cycles and ovulation, and stimulates the development of the breasts.

The *menstrual cycle* (see p. 218) is regulated by the cyclical production of oestrogens and progesterone.

Clinical features
Premenstrual tension may be due to sodium and water retention produced by oestrogens. *Contraceptive pills* contain two synthetic hormones—an oestrogen and a progestogen. These stop the ovaries releasing ova, stop endometrial development, and cause the mucus of the cervix to become so thick that sperm cannot pass up it.

PLACENTA

The placenta (see p. 312–13) secretes several hormones including *chorionic gonadotrophin* which has a lutenizing action, i.e. it develops and maintains the corpus luteum in the ovary during pregnancy.

Clinical features
Pregnancy tests are based on the detection of human chorionic gonadotrophin (HCG) in the urine. An immunological test requires only about one hour to be carried out.

PANCREAS

The intra-alveolar cell islets, also called islets of Langerhans (Fig. 20.6), are groups of cells lying between the pancreatic juice secreting cells. They form only about 1 per cent of total pancreatic tissue. Staining methods reveal two types of cells in the islets—alpha cells and beta cells.

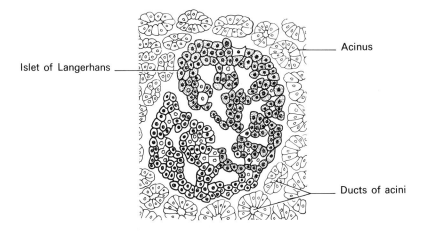

Islet of Langerhans

Acinus

Ducts of acini

Fig. 20.6. An intra-alveolar cell islet in the pancreas.

Insulin is the hormone produced in the beta cells of the intra-alveolar cell islets. It is composed of amino acids. Its production by the beta cells is stimulated by a rise in blood sugar, as occurs after a meal containing carbohydrates; it circulates in the blood and is eventually destroyed by the kidneys and liver. Its function is to stimulate the transfer of glucose across cell walls and so promote the utilization of glucose by the cells and prevent the blood sugar from rising above normal limits. *Glucagon* is the hormone produced by the alpha cells of the intra-alveolar cell islets. Its chief function is the conversion of glycogen in the liver into glucose. Its action in this way produces effects opposite to those of insulin. Its production is stimulated by a fall in the blood sugar, which can be due to fasting or to taking moderate to severe exercise.

Hormonal control of blood sugar
(normal blood sugar 4.5–8.9 mmol/l)
blood sugar lowered   *blood sugar raised*
  insulin   glucagon
                                adrenaline
                                cortisol
                                growth hormone

Clinical features
Diabetes mellitus is a disease in which there is a shortage or absence of insulin due to a failure of production; the cause of the failure is not known. The blood sugar rises, carbohydrate, fat and protein metabolism is disturbed. The administration of insulin by injection is one way of treating it. Insulin cannot be given by mouth for the amino acids of which it is composed would be subject to normal digestion in the stomach and intestine and it would be rendered useless.

KIDNEY

In addition to its functions as a controller of water balance and the chemistry of the body, the kidney produces renin and erythropoietin.

Renin is produced by certain cells in the walls of the arterioles through which the blood passes to the glomeruli. It is secreted when the blood pressure falls so low that an insufficient amount of blood passes into the kidneys. It raises the blood pressure.

Erythropoietin is secreted by the kidney in response to a reduction of the normal oxygen tension. It stimulates the formation of red blood cells in the bone marrow and increases the number available for the transport of oxygen.

The last stage in the production of biologically active vitamin D is carried out by the kidney (see p. 193).

Clinical features
Excessive secretion of renin is a factor in the production of malignant hypertension and in hypertension due to stenosis (narrowing) of a renal artery. It can cause an excessive production of androgens from the adrenal cortex and so produce excessive masculinity.

ALIMENTARY TRACT

Gastrin is a hormone produced by certain cells in the mucous membrane of the stomach. It is secreted into the blood in response to a fall in the concentration of acid in the stomach; an increase in acid reduces its secretion. Its functions are: (a) the stimulation of acid secretion by the stomach, (b) the maintenance of the tone and competence of the oesophageal sphincter where the oesophagus opens into the stomach; by keeping this sphincter tight it prevents the stomach contents from getting into the oesophagus.

Secretin is a hormone produced by cells of the duodenum on the entry of acid into the duodenum from the stomach. It stimulates the secretion of pancreatic juice.

Cholecystokinin-pancreozymin (produced by cells of the small intestine on the entry of food) stimulates the secretion of pancreatic juice and causes contraction of the gall-bladder.

PINEAL GLAND

The pineal gland is a tiny pink oval structure lying in the groove between the two superior quadrigeminal bodies of the midbrain. It is composed of epithelial cells.

The pineal gland produces *melatonin*, which appears to have a delaying action on the onset of puberty.

Clinical features
Tumours pressing on the pineal gland can reduce the production of melatonin and so cause precocious puberty. In adult life it becomes calcified and visible on an X-ray. It should lie exactly in the midline of an anterior-posterior (AP) X-ray of the skull, and a deviation of it to one side or the other is evidence of the presence of a tumour or other space-occupying lesion which is pushing it over.

21

The Nervous System

Nervous tissue is composed of:

neurones: nerve cells and their fibres,

neuroglia: cells whose functions are not fully known; some are phagocytic, absorbing and destroying micro-organisms and foreign substances that have got into nervous tissue.

NEURONE

A *neurone* (Fig. 21.1) is the basic unit of the nervous system. There are many millions of them in the nervous system. Each consists of a nerve cell and its fibres. Nerve cells vary in shape and size according to their different functions. Each cell has a nucleus and a number of granules and fibrils in its cytoplasm. *Dendrites* are the short brush-like fibres attached to the outside of a cell; through them impulses enter the cell from other cells. The *axon* is the fibre through which nerve impulses leave the cell to be transmitted to other cells. Each nerve cell has one axon, which can vary in length from a few millimetres to many centimetres. An axon often branches extensively near its end, and each branch ends in a button-like enlargement, which is its information-delivering part. A neurone never divides or is replaced; many of them die and disappear every year from birth onwards, and if a man were to survive to 80 years he is likely to have lost 1 per cent of the neurones he was born with.

Each fibre is enclosed within a thin sheath. In addition most axons are myelinated, i.e. enclosed within a sheath of *myelin*, a lipid substance. Myelination of fibres begins at the 6th month of fetal life and is completed after birth. It is probable that a fibre cannot function properly until it is myelinated. The myelin coat is not continuous along the length of a fibre, for at the *nodes of Ranvier* (Fig. 21.2), at intervals of about 1 mm along the fibre, there is no myelin and the fibre is enclosed only within its other sheath.

The transmission of a nervous impulse occurs only in one direction: into a cell through its dendrites and out of it through the axon. The cell body receives incoming signals through its dendrites, combines and integrates them, and sends other signals along the axon. One neurone receives signals from hundreds or thousands of other neurones and feeds them into hundreds or thousands of others. The signalling system is both chemical and electrical.

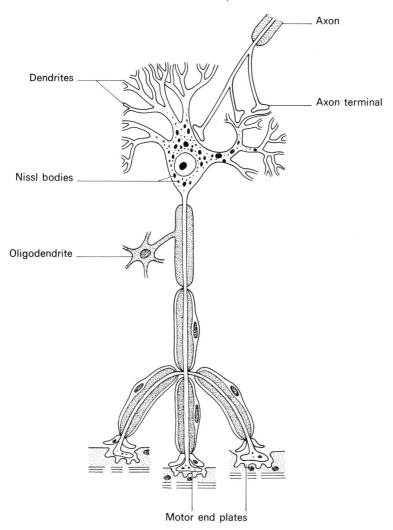

Fig. 21.1. Essential structures of a lower motor neurone.

The *nerve impulse* is a complex chemico-electrical change which travels along a nerve fibre. In it, ions (charged particles) move from the inside of an axon to the outside, and others move from the outside to the inside. As a wave passes along an axon potassium ions (K^+) leave the axon and sodium ions (Na^+) go in. The movement of one K^+ ion seems to stimulate the movement of the one next to it and so on along the axon. The nerve impulse is the result of the difference in electrical potential between K^+ and Na^+. After the wave has passed along the axon, the K^+ ions and Na^+ ions return relatively slowly to their original position.

Immediately after the passage of the nerve impulse there is a brief period during which another impulse cannot pass along the fibre.

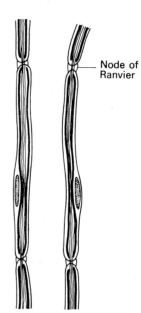

Node of
Ranvier

Fig. 21.2. Nodes of
Ranvier in a medullated
nerve fibre.

A *synapse* is the point of communication between one neurone and another. At this point the transmission of the impulse is done chemically. On the arrival of the impulse at a synapse, a chemical transmitter is liberated and stimulates the next cell. About 30 transmitters are known, among them acetylcholine, nor-epinephrine and dopamine, and each is probably concerned with a different kind of nervous system activity. Interference with them is thought to cause mental or physical disease of the brain.

Clinical features
Demyelinating diseases are those in which the myelin sheath degenerates and is absorbed. Multiple sclerosis is the commonest of them.

THE PARTS OF THE NERVOUS SYSTEM

The parts of the nervous system are:
  the central nervous system: brain and spinal cord;
  peripheral nervous system: cranial and spinal nerves,
  autonomic nervous system: sympathetic and parasympathetic systems.
  Grey matter is nervous tissue in which there is a preponderance of nerve cells. *White matter* is nervous tissue in which there is a preponderance of nerve fibres.

The brain

> *Parts of the brain*
>   cerebral hemispheres, right and left
>   midbrain
>   pons
>   cerebellum
>   medulla oblongata (continuous with spinal cord)

THE CEREBRAL HEMISPHERES

The cerebral hemispheres (Fig. 21.3, 21.4) are the largest parts of the brain. Each consists of:

a cortex, an outer layer of nerve cells, arranged in layers; it is about 2 mm thick and contains about 70 per cent of all neurones in the nervous system.

nerve fibres, running to and from these cells, connecting parts of the brain and connecting the brain with the spinal cord,

thalamus, *basal ganglia* and other masses of nerve cells within the cerebral hemisphere.

The *corpus callosum* is a thick band of fibres connecting the cerebral hemispheres; through it sensory information is exchanged between the hemispheres.

The surface of the cerebral hemispheres is marked by gyri (ridges) and sulci (fissures); about four-fifths of the total cortex is not visible on the surface, being hidden in the culci. The shape and number of the gyri vary a little in different people. The important sulci are:

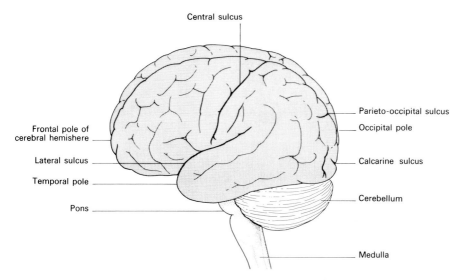

Central sulcus

Parieto-occipital sulcus

Frontal pole of cerebral hemishere

Occipital pole

Lateral sulcus

Calcarine sulcus

Temporal pole

Cerebellum

Pons

Medulla

Fig. 21.3 The brain seen from the side.

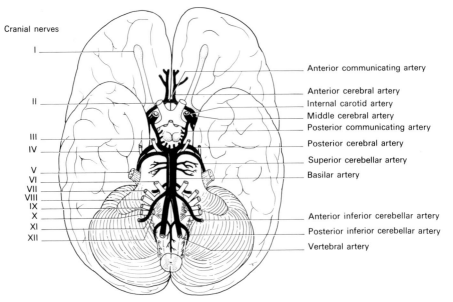

Fig. 21.4. The inferior surface of the brain, the cranial nerves, and the blood vessels on the inferior surface.

central sulcus: runs downwards and slightly forwards from about the middle of the top of the brain,

lateral sulcus: comes round from the base of the brain to run backwards and slightly upwards on the outer surface,

parieto-occipital sulcus: cuts the upper border of the brain a little way from the posterior tip.

Each cerebral hemisphere is divided into 4 lobes:

frontal lobe: in front of the central sulcus,

parietal lobe: between the central sulcus and the parieto-occipital sulcus,

occipital lobe: behind the parieto-occipital sulcus,

temporal lobe: below the lateral sulcus.

The *frontal lobe* shows:

(a) precentral gyrus: a gyrus immediately in front of the central sulcus; this is the motor area of the brain, in which are situated many (but not all) of the nerve cells in which muscular movements are stimulated.

(b) premotor cortex: the rest of the frontal lobe in front of the precentral gyrus; this is one of the 'association areas', responsible for thought processes; it also contains some cells of motor fibres.

The *parietal lobe* shows:

(a) postcentral gyrus: the gyrus immediately behind the central sulcus; this is the sensory area of the brain in which sensations of touch, pressure and slight changes of temperature are appreciated.

(b) a number of gyri behind the postcentral gyrus; they form another association area.

The *occipital lobe* contains the visual area of the brain, to which are referred sensations coming from the eyes.

The *temporal lobe* contains the auditory area, to which are referred sensations coming from the ear.

The *thalamus* is a large egg-shaped mass of nerve cells lying within the white matter. The *basal ganglia* are the lenticular nucleus, the caudate nucleus and some smaller ganglia, all lying within the white matter.

The *speech area* is the part of the cortex concerned with all aspects of speech (hearing, speaking, reading, writing). It is situated in the left hemisphere in all right-handed people and most left-handed. It includes adjacent lower parts of the frontal and parietal lobes and the upper part of the temporal lobe.

Clinical features
Aphasia is a defect of speech in which there is an inability to produce speech, spoken or written, or to understand it when spoken or written. It is due to disease or injury of the speech area.

THE MIDBRAIN

The midbrain is a small structure between the cerebral hemispheres above and the pons below. It is mainly composed of nerve fibres passing up and down it. The corpora quadrigemina are four small rounded masses of nerve cells on its dorsal surface.

THE PONS

The pons (Fig. 21.5) is a thick mass of nerve tissue continuous with the midbrain above and the medulla oblongata below. It is composed mainly of nerve fibres passing upwards or downwards or transversely into the cerebellum.

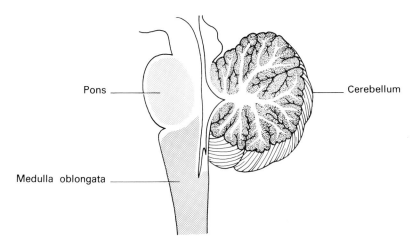

Pons _____

Cerebellum

Medulla oblongata _____

Fig. 21.5. The hind brain.

THE MEDULLA OBLONGATA

The medulla oblongata (Fig. 21.5) is a narrow piece of nervous tissue continuous with the pons above and the spinal cord below. It is composed mainly of nerve fibres. It contains the cardiac and respiratory centres of cells through which the heart and lungs are controlled.

The *brain stem* is the midbrain, pons and medulla oblongata considered as a functioning unit. The *reticular system* is a system of nerve cells and their connecting fibres within the brain stem. It is connected with all ascending and descending tracts and with all other parts of the central nervous system. It acts as an integrator of the entire nervous system. It is involved in sleep and consciousness, temperature regulation, gastrointestinal movements, respiration, circulation and metabolism.

THE CEREBELLUM

The cerebellum (Fig. 21.5) is composed of a small central lobe and larger right and left lobes. It is connected by nerve fibres (in bundles called peduncles) with the midbrain, pons and medulla oblongata. It lies below the occipital lobes of the cerebral hemispheres, being separated from them by a shelf of dura mater. Its surface is marked by a number of thin parallel gyri. Its grey matter is on its surface and in some small nuclei inside.

The chief functions of the cerebellum are the coordination of muscular activity, the control of muscle tone and the maintenance of posture and equilibrium. How it carries out these functions is unknown.

Clinical features
Diseases of the cerebellum (such as tumours, abscesses, or degenerating conditions) are likely to produce disturbances of balance, posture, muscle tone and gait.

THE VENTRICLES

The ventricles (Fig. 21.6) are a series of interconnected chambers within the brain. They are:
    lateral ventricles: one in each cerebral hemisphere, extending into the frontal, occipital and temporal lobes,
    3rd ventricle: a narrow chamber in the midbrain, communicating with the lateral ventricles above and, through a narrow tube called the aqueduct, with the 4th ventricle below.
    4th ventricle: a lozenge-shaped space lying between the pons and medulla oblongata in front and the cerebellum behind.

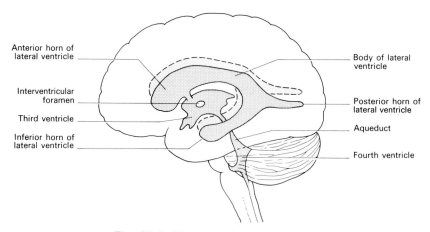

Fig. 21.6. The ventricles of the brain.

THE CEREBROSPINAL FLUID

The cerebrospinal fluid (CSF) occupies the ventricles. It is a clear fluid formed from the blood plasma in the choroid plexuses. The choroid plexuses are tangled whorls of capillaries lying within the ventricles, the largest of them being in the lateral ventricles, where most of the CSF is formed. About 500 ml are secreted daily.

Cerebrospinal fluid: normal values		
pressure	70–160 mm water	
volume	120–140 ml	
protein	20–45 mg/100 ml	20–45 g/l
glucose	50–85 mg/100 ml	2.2–3.4 mmol/l
chloride	120–130 mEq/litre	120–130 mmol/l
cells	0–5 lymphocytes/mm³	

The CSF passes through the ventricles and leaves the 4th ventricle through three small holes in its roof to enter the subarachnoid space around the brain

Functions of cerebrospinal fluid
to maintain a constant volume within the skull by increasing or decreasing in amount with any decrease or increase of the other cranial contents
to act as a buffer protecting the brain from any jar
to receive waste products of metabolism in the brain and to transfer them to the blood

and spinal cord. It is eventually absorbed into the blood, most (it is thought) by passing through the arachnoid villi, which are small vascular projections from the large venous sinuses within the skull.

The *blood-brain barrier* is the 'barrier' in the choroid plexuses which prevents certain substances (including some drugs) from passing from the blood into the CSF in the proportions that would be expected. The cells of the choroid plexuses seem to have the ability either to prevent their passage or to destroy them. The barrier protects neurones from attack by chemical substances which might harm them.

Clinical features
Specimens of CSF are obtained by the operation of lumbar puncture in which a needle is inserted into the subarachnoid space between the 2nd and 3rd lumbar vertebrae, i.e. below the level of the spinal cord, which is thus not in danger of being punctured.

  Hydrocephalus is an enlargement of the ventricles due to an obstruction to its flow at either the aqueduct between the lateral and 3rd ventricles or in the roof of the 4th ventricle; common causes are a congenital deformity, meningitis, tumour. A *meningitis* produces an increase in the number of cells and the amount of protein in the CSF. A *cerebral haemorrhage* will cause blood to be present in the CSF. *Increase in CSF pressure* can be due to intracranial tumour, intracranial haemorrhage, hypertension, meningitis, encephalitis, hydro-cephalus, and will cause severe headache and vomiting.

The spinal cord

The spinal cord (Fig. 21.7) is continuous above with the medulla oblongata. It is about 45 cm long, occupies the upper two-thirds of the vertebral canal, and ends at the level of the 1st or 2nd lumbar vertebra by tapering into a cone. This cone is connected to the coccyx by the filum terminale, a strand of connective tissue

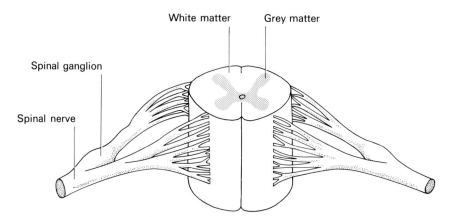

White matter   Grey matter

Spinal ganglion

Spinal nerve

Fig. 21.7. The anterior and posterior branches of a spinal nerve and their attachments to a segment of spinal cord. The posterior branch has a ganglion in it.

enclosed in the meninges. It is composed of nerve fibres on the outside (white matter) and an H-shaped mass of nerve cells (grey matter) in the middle (Fig. 21.8). A small central canal runs through the grey matter.

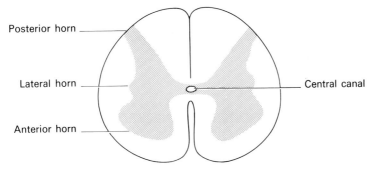

Fig. 21.8. Section through the spinal cord, showing white and grey matter.

The *nerve fibres* are arranged in three groups: the anterior, lateral and posterior columns. They consist of:
(a) motor fibres running downwards in the lateral and anterior columns,
(b) sensory fibres running upwards in the lateral and posterior columns,
(c) short fibres interconnecting different levels of the cord.

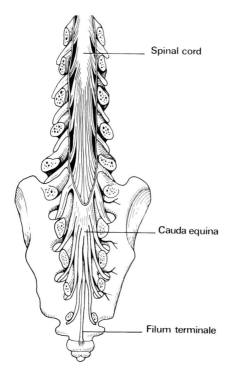

Fig. 21.9. The lower end of the spinal cord and the cauda equina.

The *nerve cells* are arranged in three main groups:
(a) an anterior horn of motor cells,
(b) a posterior horn of sensory cells,
(c) a lateral horn of sympathetic cells.

Spinal nerves are attached by anterior and posterior roots to the whole length of the cord. The spinal cord is shorter than the vertebral canal. To reach the right hole through which they have to pass the spinal nerves have to run a more and more oblique course. The lower fibres form a cauda equina (horse's tail) within the canal (Fig. 21.9).

The meninges

The meninges are the coverings of the brain and spinal cord. They consist of the following parts:

Dura mater: a thick rough white membrane enclosing the whole of the brain and spinal cord. It is composed of two layers fused together except within the skull where the outer layer is attached to the bone and where there are venous sinuses. The *falx cerebri* is a vertical fold of dura mater separating the two cerebral hemispheres in the midline. The *tentorium* is a horizontal shelf of dura mater separating the occipital lobes from the cerebellum.

Arachnoid membrane: a delicate membrane fused in places with the pia mater, in others separated from it by a *subarachnoid space*, which is filled with CSF. The cisterna magna is a large part of the subarachnoid space at the back of the hindbrain, occupying the gap between the cerebellum and the medulla oblongata.

Pia mater: a fine membrane attached to the surface of the brain and spinal cord throughout. *Leptomeninges* is a term used to describe the pia and arachnoid considered as one.

Clinical features
Meningitis is an infection of the arachnoid membrane and pia mater and can be caused by various micro-organisms. The meninges become inflamed and swollen; the cranial nerves can become involved because they pass through the subarachnoid space, involvement of blood vessels can reduce the supply of blood to the cortex and start degenerative changes. An *extradural haemorrhage* is one between the skull and the dura mater. A *subdural haemorrhage* is one between dura mater and brain. A *meningioma* is a tumour arising from the dura mater.

The sensory system

The central nervous system is continuously fed with information about the world outside the body and about the state of the organs and tissues within the body.

> *Sources of sensation*
>   organs of special sense
>   skin
>   muscles, tendons
>   internal organs

Stimuli from the skin, muscles and tendons are received on specialized sensory organs and then passed along the cranial and spinal nerves to the central nervous system.

From the trunk and limbs sensory impulses pass along the sensory nerve fibres towards the spinal cord. Just before reaching the spinal cord the sensory fibres come together to form a posterior nerve root. The ganglion of the posterior nerve root is a small swelling on it in which lie the cells of the sensory fibres. Sensory impulses pass thence into the cord. Within the cord some fibres synapse with the cells of motor nerves at the same level. Others pass upwards towards the brain, crossing at various levels to the opposite side of the cord or brain stem; impulses from one side of the body are therefore appreciated in the other side of the brain.

Clinical features

Syringomyelia is a degeneration of the grey matter of the spinal cord in its cervical part. It is likely to involve the fibres transmitting sensations of pain and temperature as they pass from one side of the spinal cord to the other; the patient cannot feel pain, heat or cold in his hands and is likely to burn himself without knowing it. *Tabes dorsalis* is a syphilitic degeneration of the posterior nerve roots and the sensory tracts in the posterior columns of the spinal cord; it can produce attacks of pain and anaesthesia of the skin.

THE APPRECIATION OF SENSATIONS IN THE BRAIN

The main areas for the appreciation of sensations are the thalamus and the sensory area.

The *thalamus* is a relay station for sensory stimuli. It receives impulses coming up to it from the spinal cord and from the cerebellum and refers some of them to the sensory area in the parietal lobe and to other lobes. It communicates with the hypothalamus which lies immediately below it. Its main functions are:
(a) the appreciation of sensations of severe heat, cold and pain,
(b) the sorting out and selection of those sensory impulses which have to be routed to the cortex.

The *sensory area*, to which fibres pass from the thalamus, is in the postcentral gyrus of the parietal lobe. Except severe degrees of heat, cold and pain, sensory impulses are appreciated there. In it (as in the motor area of the brain) the body is represented upside down—with a large area at the bottom for the head, above that a large area for the hand, and then smaller areas for the arm, trunk, leg, foot and perineum. As sensory fibres cross in the central nervous system from one side to the other, the parietal cortex of one side appreciates stimuli coming from the other side of the body.

Proprioceptive stimuli are sensory stimuli arising in the muscles and tendons; the information given is that of the degree of stretch in them. *Enteroceptive stimuli* are sensory stimuli arising in internal organs. Both these kinds of stimuli are fed into the central nervous system without coming into consciousness and are acted upon automatically. The sensory nerve endings in internal organs respond to stretch and to shortage of oxygen, but not to cutting and burning. The pain impulse arising in an internal organ may be 'felt' in an area of skin supplied by the same nerve level; hence one gets 'referred pain'.

Clinical features

Diseases of the thalamus usually cause exaggeration of pain and poor localization of pain and temperature; some lesions cause intellectual deterioration, personality changes and ataxia.

Diseases of the sensory area of cortex can produce:
(a) impairment of appreciation of touch and of ability to distinguish shape and size,
(b) epileptic sensory phenomena of abnormal sensations in the other side, which can happen when a disease process irritates the cortex.

The motor system

The motor system (Fig. 21.10) consists essentially of:
    upper motor neurones—from the cerebral cortex to the brain stem and spinal cord,
    lower motor neurones—from the brain stem and spinal cord to the muscles.
    The *upper motor neurones* arise in the motor area of the precentral gyrus and in other parts of the cortex (Fig. 21.11), especially the premotor area of the frontal lobe. In the precentral gyrus the parts of the body are represented upside down—with a large area for the head at the bottom, a large area for the hand above that for the head, and then smaller areas for the arm, trunk, leg and perineum. The more precise the movements of a part, the greater the amount of cortex responsible for it.
    The upper motor neurones form the *pyramidal tract*. It is composed of corticonuclear fibres which run only as far as the brain stem, for their connections are with those cranial nerves which have motor functions, and corticospinal fibres which run to the spinal cord. The pyramidal tract runs downwards and inwards through the cerebral hemisphere, and then through the midbrain, pons and medulla oblongata, forming in the medulla a long ridge, the pyramid, from which it gets its name. In the medulla most of the fibres cross to the other side and run down the anterior column of the spinal cord. Those which do not cross in the medulla run down the anterior column, but they too cross in time. Accordingly one side of the brain directs and controls movements on the other

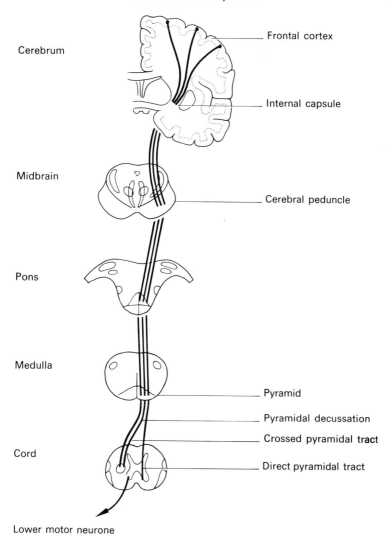

Cerebrum

Frontal cortex

Internal capsule

Midbrain

Cerebral peduncle

Pons

Medulla

Pyramid

Pyramidal decussation

Crossed pyramidal tract

Cord

Direct pyramidal tract

Lower motor neurone

Fig. 21.10. The course of the motor (pyramidal) fibres.

side of the body. (No one has the slightest idea why nervous pathways should cross from one side to the other as they do.)

In the spinal cord the motor fibres end by synapsing with motor cells in the anterior horn of grey matter.

The *lower motor neurones* have their cells in the brain stem for cranial nerves and in the anterior horn of grey matter for the spinal nerves. The motor fibres leave the spinal cord in the anterior root of a spinal nerve. This nerve divides into a small branch for the muscles of the back and a large anterior branch for the muscles at the sides and front of the body and to the muscles of the limbs. As it approaches a muscle, the nerve fibre loses its myelin sheath and divides into a

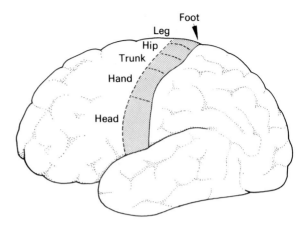

Fig. 21.11. The motor
cortex of the left cerebral
hemisphere.

number of branches. The ending of a nerve fibre on a muscle is in a *motor end plate*, a flat disc lying on a muscle fibre. Impulses are transmitted chemically from nerve to muscle.

Effects of upper and lower motor neurone lesions	
Upper motor neurone	*Lower motor neurone*
some muscle weakness	muscle weakness
no muscle wasting (except after long disuse of part)	muscle wasting
muscle tone increased	muscle tone decreased
tendon reflexes increased	tendon reflexes decreased
plantar reflex extensor (upgoing big toe)	plantar reflex flexor (downgoing big toe)
growth of part unaffected (in childhood)	growth of part diminished (in childhood)

THE EXTRAPYRAMIDAL SYSTEM

The *extrapyramidal system* is a motor system other than the pyramidal system. It is composed of a number of cells in the cerebral cortex, the basal ganglia and various small nuclei in the cerebral hemispheres, brain stem and cerebellum, and of the fibres which connect these cells with one another and with motor cells in the brain stem and spinal cord.

The functions of the extrapyramidal system are the control of coarse movements, associated movements (such as swinging the arms when walking), postural adjustments and muscle tone.

Clinical features
Damage to the extrapyramidal system can cause spasticity of muscles, tremor and slowness of voluntary movements, as seen in parkinsonism. Athetosis— involuntary twisting movements—can occur in some diseases.

The hypothalamus

The *hypothalamus* is a small central area of nerve cells just below the thalami. It is connected with the thalami, with the upper end of the autonomic nervous system, and with the pituitary gland, which is just below it and connected to it by a stalk.

The hypothalamus is an important centre for the integration of basic functions for an individual. It is also part of the endocrine system, for it is closely linked with the pituitary gland, feeding it with chemical factors which pass down the pituitary stalk into the gland and control its hormonal activity.

Functions of hypothalamus
control of biological clock, which regulates 24-
hourly activities—sleep, temperature,
hormone secretion
control of appetite
control of water balance
integration of emotional reactions.

The cranial nerves

The cranial nerves are 12 pairs of nerves whose central connections are in the brain.

Cranial nerves
1. olfactory nerve
2. optic nerve
3. oculomotor nerve
4. trochlear nerve
5. trigeminal nerve
6. abducent nerve
7. facial nerve
8. auditory nerve
9. glossopharyngeal nerve
10. vagus nerve
11. accessory nerve
12. hypoglossal nerve

OLFACTORY NERVE

The olfactory nerve consists of a number of short fibres which run from the olfactory area at the top of the nasal cavity through small holes in the cribriform plate of the ethmoid bone to enter the skull, and end in a tract which transmits the impulses to the smell area of the brain.

Clinical feature
A fracture of the front of the base of the skull may cut the nerves or tract and cause loss of smell. A tumour of the frontal lobe can cause loss of smell by pressing on the tract.

OPTIC NERVE

The cells of the optic nerve are in the retina. Each optic nerve contains about a million fibres, each connected with a rod or cone in the retina. The nerves pass backwards through the optic foramen at the back of the orbit and unite with each other in the optic chiasma. The *optic chiasma* is a cross in which the fibres from the inner half of each retina pass to the opposite side, while those from the outer half stay on the same side. From the chiasma the optic tract on each side runs backwards and through various connections the visual impulses are transmitted to the visual area of the brain in the occipital lobe.

Clinical features
Diseases or injuries of the optic nerve cause impairment of vision or blindness. Tumours of the pituitary gland can cause blindness in the outer fields of vision by pressing on the fibres crossing in the optic chiasma, which is just above and in front of the gland.

OCULOMOTOR NERVE; TROCHLEAR NERVE; ABDUCENT NERVE

The oculomotor nerve, trochlear nerve and abducent nerve are the nerves of the muscles which move the eyeball. The 3rd nerve supplies all the muscles except the superior oblique (4th cranial nerve) and the lateral rectus (6th cranial nerve). They enter the orbit through the superior orbital foramen.

Clinical features
The nerves are liable to be involved in a meningitis during their relatively long course through the subarachnoid space.

TRIGEMINAL NERVE

The trigeminal nerve (Fig. 21.12) combines a large sensory part and a small motor part.

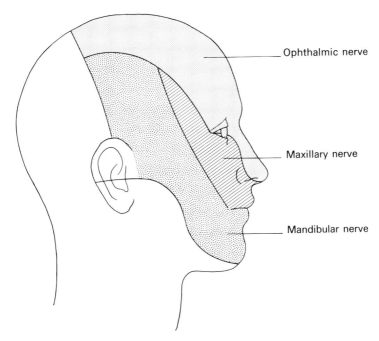

Ophthalmic nerve

Maxillary nerve

Mandibular nerve

Fig. 21.12. The distribution of the sensory branches of the trigeminal (5th cranial) nerve.

The *sensory part* is the sensory nerve for the face and scalp. It has three branches:

  ophthalmic nerve: from forehead, upper eyelid, conjunctiva, top of nose,

  maxillary nerve: from cheek, upper jaw, maxillary sinus,

  mandibular nerve: from lower jaw.

The *motor branch* supplies the muscles of mastication (masseter, temporalis and the pterygoids).

Clinical feature

Trigeminal neuralgia is a disease, of unknown origin, characterized by attacks of severe pain in the region supplied by one of the branches of the trigeminal nerve.

FACIAL NERVE

The facial nerve (Fig. 21.13) is the motor nerve to the muscles of facial expression. It passes through a narrow canal in the temporal bone to emerge through a small hole at the base of the skull just behind the parotid gland. It divides into branches which run through the gland to get to the muscles of the face.

Clinical feature

Bell's palsy is a paralysis of the facial muscles on one side of the face, possibly as a result of the compression of the nerve by swelling as it passes through the canal in the temporal bone.

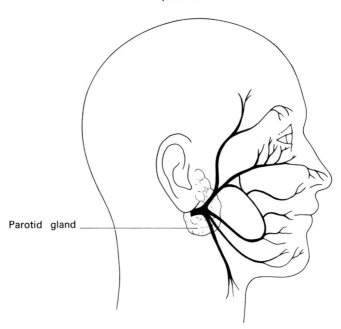

Parotid gland

Fig. 21.13. The distribution of the facial (7th cranial) nerve to the muscles of the face.

AUDITORY NERVE

The auditory nerve runs from the brain to the inner ear inside the petrous portion of the temporal bone. It has two parts:
    the *cochlear nerve*: the nerve of hearing,
    the *vestibular nerve*: the nerve of balance and position in space.

Clinical features
Diseases of the cochlear nerve produces deafness. Diseases of the vestibular nerve produce giddiness (vertigo) and impairment of balance.

GLOSSOPHARYNGEAL NERVE

The glossopharyngeal nerve supplies sensory fibres to the pharynx and back of the tongue.

VAGUS NERVE

The vagus is a long nerve (Fig. 21.14) which passes down the neck and thorax into the abdomen. It is essentially a transmitter of the impulses of the parasympathetic system. It has many branches through which it supplies the larynx, pharynx, heart, lungs, kidney, liver, spleen, and the alimentary tract as far as the descending colon.

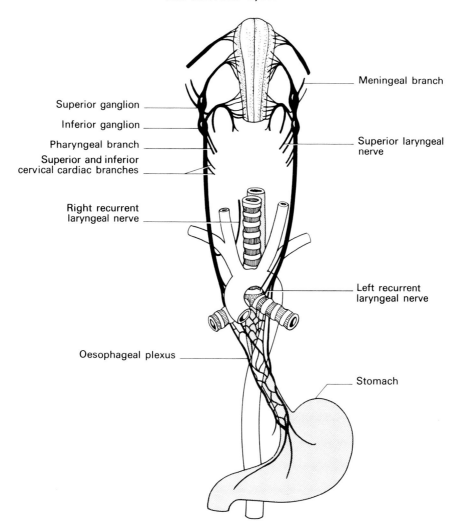

Superior ganglion
Inferior ganglion
Pharyngeal branch
Superior and inferior cervical cardiac branches

Right recurrent laryngeal nerve

Oesophageal plexus

Meningeal branch

Superior laryngeal nerve

Left recurrent laryngeal nerve

Stomach

Fig. 21.14. The distribution of the vagus (10th cranial) nerve.

ACCESSORY NERVE

The accessory nerve runs diagonally across the neck to supply the sternomastoid and trapezius muscles.

HYPOGLOSSAL NERVE

The hypoglossal nerve is the motor nerve to the muscles of the tongue.

Clinical feature

Paralysis causes wasting of the tongue on that side and deviation to the same side when the tongue is put out, owing to the unopposed action of the muscle on the other side.

The spinal nerves

```
Spinal nerves: 31 pairs
    cervical          8 (C1–8)
    thoracic         12 (T1–12)
    lumbar            5 (L1–5)
    sacral            5 (S1–5)
    coccygeal         1 (Co 1)
```

Each nerve is formed by the union of:
an anterior (motor) root,
a posterior (sensory) root,
from each segment of the spinal cord. There is a ganglion (of nerve cells) on the posterior root.

Each nerve divides into 2 branches, in each of which there are motor and sensory fibres: (a) a small posterior branch to the skin and muscles at the back of the body, and (b) a larger anterior branch to the skin and muscles at the side and front of the body and in the limbs.

A pair of spinal nerves supplies one segment of the body from one segment of the spinal cord. The simple pattern is best seen in the thoracic region, and has elsewhere become altered by the development, from the segments, of the limbs. Because of the way in which the limbs have developed, the anterior branches have formed plexuses, which are essentially regroupings of the motor and sensory fibres into nerves which pass to appropriate muscles and areas of skin.

The thoracic nerves show the simple segmentation pattern, except for the anterior branch of the 1st which joins the brachial plexus.

The *posterior branches* of these nerves supply the muscles and skin at the back of the chest.

The larger *anterior branches* run in the intercostal spaces, give off a lateral branch to the side of the body and an anterior branch at the front, supplying motor branches to the intercostal muscles and sensory branches to the skin.

The course of these nerves is obliquely downwards in the line of the ribs; and T7–12 supply the muscles of the anterior abdominal wall and the skin of the front of the abdomen.

THE CERVICAL PLEXUS

The cervical plexus is a plexus in the neck formed of the anterior branches of C1–5. From it arise:
(a) nerves to the muscles and skin of the neck and shoulders,
(b) the *phrenic nerve* (C3–5), which runs down the neck and thorax to supply the diaphragm.

THE BRACHIAL PLEXUS

The brachial plexus (Fig. 21.15) is a plexus in the lower part of the neck and axilla formed of the anterior branches of C5-8, T1. From it arise:

pectoral nerves: to the pectoral muscles in the front of the chest,

circumflex nerve: to the deltoid muscle, shoulder joint, and skin over the shoulder,

musculocutaneous nerve: to the biceps and other muscles and to the skin on the outer side of the forearm,

radial nerve: to the triceps, brachioradialis and extensor muscles of the forearm, and to the skin on the outer side of the arm and the back of the forearm,

median nerve: to most of the flexor muscles of the forearm and to many of the small muscles of the hand, and to the skin on the lateral side of the hand,

ulnar nerve: to muscles of the forearm and hand, and to the skin on the medial side of the forearm and hand.

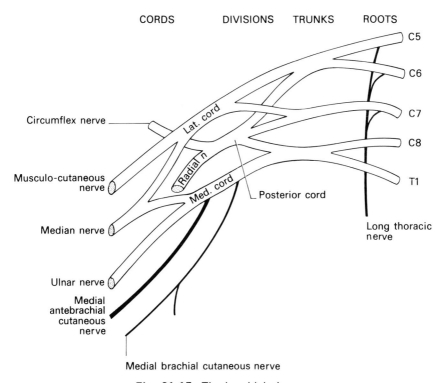

Fig. 21.15. The brachial plexus.

The lumbar plexus is formed by anterior branches of T12, L1–4. From it arise:
  femoral nerve: to the quadriceps muscle and other muscles in the front of the thigh,
  obturator nerve: to the adductor muscles on the inner side of the thigh.

THE SACRAL PLEXUS

The sacral plexus is formed in front of the sacrum and behind the rectum by the anterior branches of L4–5, S1–4. From it arises the *sciatic nerve* (Fig. 21.16), which runs deep in the buttock and back of the thigh and divides in the popliteal fossa into medial and lateral popliteal nerves; it supplies the hamstring muscles at the back of the thigh and all the muscles below the knee.

Clinical features
Neuropathy (neuritis) is a degeneration of spinal nerves and their branches as a result of infection, alcoholism, diabetes, poisoning by arsenic and other substances etc. *Sciatica* is due to pressure on lumbar and sacral nerve roots within the vertebral canal by a slipped disc (a protrusion of the nucleus pulposus through the annulus fibrosus of an intervertebral disc).

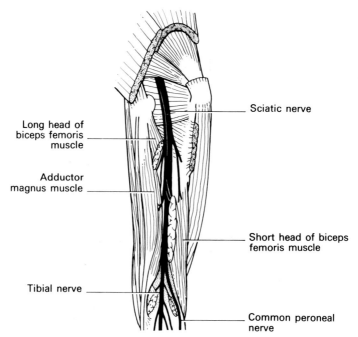

Fig. 21.16. The sciatic nerve in the back of the thigh.

TENDON REFLEXES

A tendon reflex is a muscular contraction produced in response to the stretching of a muscle.

```
┌─────────────────────────────────────────────────┐
│ Tendon reflexes tested in clinical examination    │
│      head:      jaw jerk                          │
│      arm:       biceps jerk                        │
│                 triceps jerk                       │
│                 supinator jerk                     │
│      leg:       knee jerk                          │
│                 ankle jerk                         │
└─────────────────────────────────────────────────┘
```

These reflexes are dependent upon the presence of a reflex arc. A *reflex arc* is composed of (a) sensory organs and nerve fibres which convey an impulse to the central nervous system and (b) motor fibres which convey the impulse to the muscle. The exchange between sensory and motor fibres is made in the brain stem for the jaw jerk and in the spinal cord for the others.

The mechanism in a typical jerk such as the knee jerk is this. Tapping the patellar tendon slightly lengthens the muscle spindles, structures sensitive to stretch and lying between muscle fibres close to their terminations. From the muscle spindles sensory impulses are transmitted from the quadriceps muscle to the spinal cord; the exchange is made, and motor impulses are sent down the motor fibres to the quadriceps, which by contracting produces the knee jerk.

Clinical features
Tendon reflexes may be (a) congenitally absent, (b) diminished by disease or injury affecting the sensory motor paths and their central exchange, e.g. neuropathy, tabes dorsalis, poliomyelitis), (c) increased by upper motor neurone lesions, when higher control is lost.

SUPERFICIAL REFLEXES

Superficial reflexes are muscular reactions to touching or stroking the skin or a mucous membrane. They are dependent upon intact sensory and motor supplies to the tissues involved.

```
┌─────────────────────────────────────────────────┐
│ Superficial reflexes tested in clinical examination│
│   conjunctival reflex: a blink stimulated by       │
│       touching the conjunctiva                     │
│   pharyngeal reflex: contraction of the pharynx    │
│       when touched.                                │
│   abdominal reflexes: contraction of muscles of    │
│       abdominal wall in response to stroking the   │
│       abdomen; the responses are variable and      │
│       difficult to obtain in obese or tense patients│
│   cremasteric reflex: contraction of cremasteric   │
│       muscle in the scrotum in response to stroking│
│       the thigh                                    │
│   plantar reflex: flexion of toes in response to   │
│       stroking the sole.                           │
└─────────────────────────────────────────────────┘
```

Chapter 21

Clinical features

An upper motor neurone lesion can produce absence of abdominal reflexes and an extensor plantar response (Babinski sign), in which the big toe turns up and the other toes fan out. An up-going toe is a normal response in babies.

The autonomic nervous system

The autonomic nervous system is composed of two parts, the sympathetic system and the parasympathetic system, which between them supply the blood vessels, the internal organs and the endocrine glands (especially the adrenal medulla). The word 'autonomic' means 'self-governing'; but the system is not completely self-governing, for its functions are integrated with the central nervous system.

THE SYMPATHETIC SYSTEM

The sympathetic system is composed of:
(a) controlling centres in the cortex, hypothalamus and medulla oblongata,
(b) cells in the lateral horn of grey matter in the spinal cord, cells which are connected with
(c) a chain of ganglia fibres running from the neck to the abdomen, lying against the bodies of the veretebrae,
(d) fibres which run from the ganglia to form complex plexuses on arteries and other organs:

The *cervical ganglia* distribute branches to the carotid and axillary arteries and their branches, and to the larynx, trachea, thyroid gland and heart.

The *thoracic ganglia* distribute branches to the brachial plexus, the heart, lungs, aorta and its branches, and abdominal organs.

The *lumbar and sacral ganglia* distribute branches to the common iliac arteries and their branches and to the pelvic organs.

THE PARASYMPATHETIC SYSTEM

The parasympathetic system is composed of:
a *cranial part*, which has connections with the cerebral cortex and the hypothalamus, and from which fibres are distributed with some of the cranial nerves (oculomotor, facial, glossopharyngeal, vagus and accessory); of these the vagus is the most important;
a *sacral part*, from which nerves pass to the organs in the pelvis.

The actions of the sympathetic and parasympathetic systems are opposed.

Functions of sympathetic and paraympathetic systems

Organ	Sympathetic	Parasympathetic
heart	rate & output increased	rate & output decreased
coronary arteries	dilated	
arteries to muscles	dilated	
arteries to organs	constricted	
BP	raised	lowered
bronchial muscle	dilated	contracted
adrenal medulla	increased secretion of adrenaline & noradrenaline	
liver	glycogen converted into glucose	glucose converted into glycogen
alimentary tract	peristalsis decreased, sphincters closed	peristalsis increased, sphincters relaxed
bladder	muscle tone decreased sphincters contracted	muscle tone increased sphincters relaxed
skin	hairs erected, sweat glands stimulated	

In emergencies the sympathetic system enables the body to cope with danger:
the heart beats faster and pumps out more blood,
the arteries dilate to provide more blood for the muscles,
more adrenaline is produced,
glucose for energy is produced by the liver,
movements of bowel and bladder are reduced.

In normal life the two systems keep bodily functions going on quietly without being brought into consciousness.

Clinical features
Sympathectomy (removal of the whole or part of the chain of sympathetic ganglia) is sometimes performed for excessive sweating, to increase the blood flow to fingers and toes threatened with gangrene, and to treat unilateral kidney disease that is causing hypertension.

Electrical activity in the brain

The simultaneous activity of millions of brain cells produces electrical discharges which are recorded by an electro-encephalogram (EEG).

The *alpha rhythm* is the normal pattern of slightly irregular small waves occurring at a rate of 8–13 a second. It begins in childhood and remains constant for an individual. The waves are best recorded when the person has his eyes shut and is relaxed. Both cerebral hemispheres produce identical waves.

Clinical features
Abnormal EEGs of much larger and more irregular waves are produced in diseases of the brain, in epilepsy, and in people with aggressive behaviour. An EEG is used to diagnose these conditions and to locate a tumour or other lesion of the brain.

Sleep

Not much is known about sleep. The parts of the brain involved appear to be the cortex, the reticular system and the hypothalamus. Sleep is composed of two parts: (a) long periods of quiet sleep, in which slow EEG waves are present and no limb or eye movements are made; (b) periods of deep sleep lasting for a few minutes and characterized by rapid EEG waves, sudden eye movements under closed lids, jerky limb movements and dreams.

Reflex actions

A *simple* (*unconditioned*) *reflex* is one in which there is an automatic response to a stimulus, such as the blinking of an eye to a bright light. These reflexes are inborn and more or less the same in all people. They are dependent upon an intact sensory and motor system. Similar reflexes occur with glands, e.g. the salivary glands secrete saliva in response to the smell and taste of food.

A *conditioned reflex* is a reflex in which the response is produced by some stimulus other than the natural one and has been built upon a natural one. Ringing a bell does not by itself produce salivation, but if the ringing of the bell immediately and always precedes the giving of food to an animal, ringing the bell will in time by itself produce salivation. Much of a person's ordinary behaviour is said to be 'conditioned' by such reflex changes.

The functions of the nervous system

The nervous system is concerned with:
(a) receiving sensations of various kinds from outside and inside the body,
(b) acting on these sensations, dealing with them automatically or feeling and thinking about them,
(c) storing memories and releasing them when required to,
(d) expressing emotions,
(e) sending messages to other parts of the nervous system, to muscles, glands, endocrine and other organs,
(f) so controlling the body that health is maintained, dangers are avoided or dealt with, and pleasurable activities promoted.

The brain is an active organ with a large blood supply and oxygen intake—it takes up about 20 per cent of all the oxygen in the blood. Neurones require glucose and oxygen, and function badly or die if the supply of either falls below the essential amount.

The size of the cerebral hemispheres is the significant feature of the human brain. The hemispheres of an individual may appear to be exact mirror-images of each other, but slight differences exist and these differences may be associated with different functions. Large areas of cortex are concerned with fairly elementary sensory and motor functions; others are concerned with specific and complicated functions (such as seeing and hearing). The left hemisphere is often called the dominant one for it controls activities of the right side of the body (about 91 per cent of people are right-handed). The left hemisphere is particularly concerned with all aspects of language. The right hemisphere appears to be more important than the left in the recognition and expression of emotion; it is possibly also involved in musical abilities.

Memory is one of the functions of the brain, but very little is known how memories and skills are stored. One theory is that they are stored in DNA. There are thought to be two kinds of memory—short term (such as remembering a telephone number for a few minutes) and long term; and each kind may be retained in a different way.

Clinical feature
Loss of memory (amnesia) occurs in diseases of the brain involving areas of the cortex, e.g. in dementia of any kind.

22

The Eye and Vision

THE ORBITS

The orbits are pyramidal cavities in the skull, with the base forwards and the apex pointing backwards and slightly medially.

The *orbital margin* is the circumference of bone on the face, being formed of a supra-orbital margin above and an infra-orbital margin below.

The principal bones forming the orbit are:

above: frontal bone,

below: maxilla,

lateral wall: zygoma, great wing of sphenoid,

medial wall: lacrimal bone, maxilla, sphenoid, ethmoid bones.

Contents of orbit
  eyeball
  optic nerve
  muscles acting on the eyeball:
    four recti muscles
    two oblique muscles
  oculomotor, trochlear, abducent nerves (III, IV, VI cranial)
  branches of ophthalmic nerve (sensory to forehead)
  blood and lymph vessels
  lacrimal gland and sac
  fat and fascia

EYEBALL

The eyeball (Fig. 22.1) is nearly spherical, being flattened slightly from above downwards. It lies cushioned in fat, protected in front by the eyelids and elsewhere by the bones of the orbit.

It is composed of:

The wall of the eye
(a) the cornea and sclera,

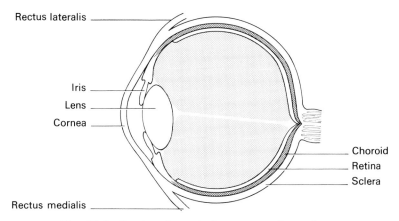

Fig. 22.1. Section through the eye seen from above.

(b) the choroid coat, ciliary body and processes, and the iris, which together form the vascular layer.

The media through which light passes
(a) the cornea,
(b) the aqueous humour in the anterior chamber,
(c) the lens,
(d) the vitreous body.

Nervous tissue
(a) the nerve cells in the retina,
(b) nerve fibres passing from these cells.

The wall of the eye

The *cornea* and *sclera* form the outer layer of the eyeball.

The *cornea* is the transparent slightly flattened dome which forms the anterior one-sixth of the wall. Around its periphery it is continuous with the sclera. It has a nerve supply but is avascular (i.e. has no blood supply).

The *sclera* is the strong opaque lateroposterior five-sixths of the wall. It appears in front as 'the white' of the eye. It is composed of fibrous tissue. At the back it is continuous with the sheath of the optic nerve.

The *choroid coat, ciliary body* and *processes*, and the *iris* have a rich blood supply and form the vascular coat through which the eye is supplied with what it needs.

The *choroid coat* is a pigmented layer between the sclera and the retina.

The *ciliary body* is composed of about 70 wrinkled ciliary processes and of a ciliary muscle. It forms a ring continuous behind the choroid body.

The *iris* is a circular, pigmented diaphragm with a slightly off-centre aperture, the *pupil.* It lies partly in front of the lens, partly in front of the ciliary body. The colour of the iris varies with the amount of pigment in it: the more pigment the darker the iris. It contains smooth muscle fibres which form: (i) the sphincter muscle of the pupil, encircling it, (ii) the dilator muscle of the pupil, with its fibres running radially away from the pupil.

The media through which light passes

(a) The *cornea*
(b) The *aqueous humour*: a fluid which occupies the anterior chamber of the eye, i.e. the space between the cornea in front and the lens and ciliary body behind. It is similar in composition to cerebrospinal fluid.
(c) The *lens*: a transparent biconvex lens with a diameter of 10 mm and a thickness of 4 mm at the centre. It lies between the iris and cornea, being separated from them by the aqueous humour. It is composed of a colloid gel enclosed in a capsule. All around the circumference of the lens this capsule is attached to the suspensory ligament of the lens. This ligament is attached at its periphery to the ciliary body.
(d) *Vitreous humour*: a gelatinous substance containing water and mucopoly-saccharide which occupies the space between the lens and the retina.

Nervous tissue

The *retina* is the nervous part of the eye, being composed of nerve cells and their fibres. It forms a layer on the inner surface of the choroid coat and is in contact with the vitreous body. A layer of pigmented cells separates the nerve cells from the choroid coat.

The nerve cells (Fig. 22.2) are
    rods: cylindrical structures numbering over 100 million,
    cones: cone-shaped structures, with the point pointing outwards numbering about 7 million.

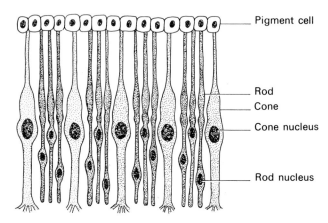

Pigment cell

Rod
Cone
Cone nucleus

Rod nucleus

Fig. 22.2. The rods and cones of the retina.

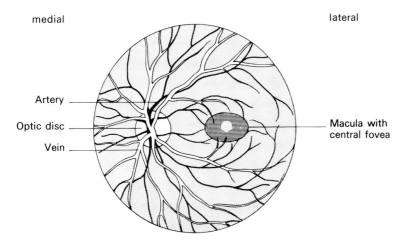

Fig. 22.3. The fundus of the eye as seen through an ophthalmoscope.

The rods and cones are packed tightly side by side. From their inner ends nerve fibres run to cells, and from these cells nerve fibres run in the innermost layer of the retina towards the place where they turn to pierce the sclera and become the optic nerve.

The *macula* (Fig. 22.3) is a yellow spot in the retina close to the posterior pole of the eye. The nerve cells in its centre are all cones.

OPTIC NERVE

The optic nerve (II cranial) is composed of about a million nerve fibres from the rods and cones of the retina.

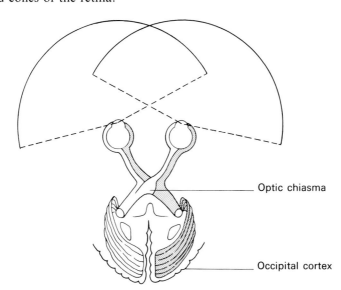

Fig. 22.4. The fields of vision and the optic nerves, chiasma and tracts.

It emerges from the eyeball slightly on the medial side of the posterior pole, passes backwards in the orbit and then goes through the optic canal in the sphenoid bone. The two optic nerves unite to form the optic chiasma. In the *optic chiasma* (Fig. 22.4) the nerve fibres from the medial (nasal) half of the retina cross to the opposite side; those from the lateral (temporal) half remain on the same side. From the chiasma the fibres run in the optic tract on each side.

The *optic tract* passes backwards to end in intermediate centres in the midbrain. Thence fibres pass to the visual centre of the brain.

The *visual centres* are situated in the occipital lobes. In each the opposite half of the field of vision is represented. The left occipital lobe receives the impulses coming from the lateral half of the left retina and the medial half of the right retina, because the fibres from the lateral half do not cross in the optic chiasma and those from the medial half do. Similarly the right occipital lobe receives the impressions coming from the other two halves.

BLOOD SUPPLY TO THE EYE

The ophthalmic artery, a branch of the internal carotid artery, enters the orbit through the optic canal with the optic nerve. It gives off:

the *central artery of the retina*, which enters the optic nerve, runs to the centre of the optic disc, and then divides into branches which radiate outwards in the retina and supply it. It is an 'end artery', i.e. it does not anastomose with any other,

ciliary arteries which pierce the sclera of the eyeball and supply blood to the choroid coat, ciliary body and iris.

VENOUS DRAINAGE OF THE EYE AND ORBIT

The veins from the structures within the orbit run backwards to the *cavernous sinus*, a venous sinus lying within the skull on either side of the sphenoid bone;

Veins from the skin of the face around the eye and nose communicate with veins in the orbit.

Clinical features
The branches of the central artery of the retina are the only arteries that can be seen in a living person. They are seen with the aid of an ophthalmoscope. Evidence of arterial disease may be found. As this artery is an 'end artery', any blocking of it, e.g. by a clot of blood, deprives the retina completely of blood.

The venous connections between the face and the cavernous sinus are important because a boil on the face can lead to infection of the veins within the orbit and thence to a dangerous infection of the cavernous sinus.

OTHER STRUCTURES

The muscles of the eyeball

Each eyeball is moved by four recti and two oblique muscles.

The *superior rectus, inferior rectus, medial rectus* and *lateral rectus* arise from a fibrous ring at the back of the orbit. They run forwards in their respective positions and are inserted into the sclera of the eyeball a little distance behind the junction of sclera and cornea.

The *superior oblique* arises at the back of the orbit, runs in the superomedial aspect of the orbit, passes through the trochlea, a fibrocartilaginous ring attached to the frontal bone, and changing direction goes backwards and outwards to be inserted into the superolateral part of the sclera.

The *inferior oblique* lies in the front of the floor of the orbit. It arises from the maxilla, runs laterally under the eyeball and then up its outer side to be attached to the sclera.

Nerve supply: the oculomotor (III cranial) nerve supplies all these muscles, except the superior oblique (IV cranial) and lateral rectus (VI cranial).

The eyeballs are moved by the muscles acting in concert. The principal actions of individual muscles are:

lateral rectus:	turns eye outwards,
medial rectus:	turns eye inwards,
superior rectus:	turns eye upwards and rotates it inwards,
inferior rectus:	turns eye downwards and rotates it inwards,
superior oblique:	turns eye downwards and rotates it inwards,
inferior oblique:	turns eye upwards and rotates it outwards.

Clinical features

A *squint* (strabismus) is an abnormal deviation of one eye so that the axes of the two eyes are no longer parallel. A deviation may be in any direction but is usually horizontal. It may be due to: (a) paralysis of one of the muscles that move the eyeball, with an unchecked action of the other muscles; or (b) a failure to develop binocular vision for any reason (such as a 'lazy eye') which prevents a child from learning to focus on an object with both eyes.

Nystagmus is a condition in which the eye oscillates, usually from side to side. It occurs in diseases of the cerebellum, in some diseases of the middle ear or auditory nerve, and in miner's nystagmus, a condition due to working in poor light.

The eyelids

Each eyelid is composed of:

a very thin *skin* on the outside,
a *tarsal plate* of dense connective tissue,

tarsal glands, modified sebaceous glands in each tarsal plate and opening by ducts on the edge of the eyelid just behind the eyelashes,

conjunctiva, a thin smooth membrane which covers the inner surface of each eyelid and is reflected on to the eyeball, where it covers the front of the cornea. The *conjunctival sac* is the space between the two layers.

eyelashes: short thick curved hairs projecting from the margin of the lid.

The *palpebral fissure* is the interval between the upper and lower eyelid. The *lateral canthus* is the outer angle of this fissure; the *medial canthus* is the inner angle.

The *orbicularis oculi muscle* is the thin circular muscle which surrounds the eye, being partly in the eyelid, partly in the face. By contracting it closes the eye. It is one of the muscles of facial expression and is supplied by the VII (facial) cranial nerve.

The *levator palpebrae superioris* (raiser of the upper eyelid) muscle arises from the back of the orbit, runs forwards at the top of the orbit and is inserted into the upper border of the tarsal plate in the upper eyelid. It is supplied by the oculomotor (III cranial) nerve.

Eyebrows are formed of fatty tissue, fibres of the orbicularis oculi muscle and hairs, and overlie the superciliary arch, a ridge on the frontal bone.

Clinical features

Conjuctivitis is inflammation of the conjunctiva. The conjunctiva becomes reddened; there may be a discharge, photophobia (dislike of light), and a feeling of grittiness due to the rubbing together of the two inflamed surfaces.

Lacrimal apparatus

The lacrimal apparatus produces and disposes of tears. It is composed of:

the lacrimal gland,

the upper and lower lacrimal ducts,

the lacrimal sac,

the nasolacrimal duct.

The *lacrimal gland* is situated in the upper and outer angle of the orbit, lying in a depression in the orbital part of the frontal bone. It is composed of secreting cells and opens by several ducts into the conjunctival sac at its superolateral angle.

The *upper and lower lacrimal ducts* are two short ducts which have an opening at the inner end of each eyelid (the opening is just big enough to be visible) and pass inwards to enter the lacrimal sac.

The *lacrimal sac*, into which the ducts open, is the blind top end of the nasolacrimal duct and is situated behind the inner canthus.

The *nasolacrimal duct* is about 2 cm long and descends through a bony canal (formed by the maxilla, lacrimal bone and inferior concha) to open into the inferior meatus of the nose, i.e. below the inferior concha.

THE TEARS

The tears are a slightly alkaline fluid produced continuously by the lacrimal gland. They enter the lacrimal sac through the ducts of the gland, keep moist the adjacent surfaces of the conjunctiva and wash away dust. Some of the fluid evaporates as it passes across the front of the eye. The rest is drawn into the lacrimal ducts by suction produced by contraction of the orbicularis oculi muscle and discharged down the nasolacrimal duct into the nose. In emotional states excess tears pour over the lower eyelid on to the face.

Clinical features
Dacryocystitis is an inflammation of the lacrimal gland. *Epiphora*, an overflowing of tears on the cheeks, is caused by dacryocystitis or to a blocking of the nasolacrimal duct.

VISION

The eye is like a camera, but works much better because it acts automatically, precisely and rapidly without any conscious adjustments having to be made.

Light

(a) enters the eye through the transparent cornea,
(b) passes through the lens, which inverts it,
(c) forms an inverted image on the retina.

The *retina* converts light into nervous impulses. These impulses pass along the optic nerve and tracts to the brain, are relayed to the occipital cortex and are there interpreted as pictures.

ENTRY OF LIGHT INTO THE EYE

The amount of light entering the eye is regulated by the size of the pupil. The iris acts as a diaphragm, the size of the pupil being controlled by its circular and radial muscle fibres.

The muscles of the iris are controlled by:
(a) sympathetic fibres which arise in the superior cervical ganglion in the sympathetic chain in the neck: impulses along them dilate the pupil by relaxing the circular fibres,
(b) parasympathetic fibres which run with the third (oculomotor) nerve: impulses along them constrict the pupil by relaxing the radial fibres.

The pupil enlarges in darkness and constricts in light. Its size at any one time is the result of a balance between sympathetic and parasympathetic stimulation.

Constriction of the pupil: (a) gives greater depth of focus because distant and near objects are in focus at the same time; and (b) reduces any distortion produced by the lens.

LIGHT REFLEX

The constriction of the pupil is the result of a light reflex. The sequence is:
   stimulation of the retina of either eye by light →
   stimuli pass along the optic nerve and tract →
   reach the superior corpora quadrigemina in the midbrain →
   are transmitted to the nuclei of the third cranial nerves on both sides →
   pass along the parasympathetic fibres →
   reach the muscles of the iris and constrict the pupil in both eyes.

Clinical features
Atropine sulphate, homatropine and other drugs are used to dilate the pupil for
diagnostic or therapeutic reasons. They act by blocking parasympathetic
activity. The pupil becomes dilated by the unopposed action of the sympathetic
nerves.

Convergence–accommodation reflex

At rest the eyes are focused on infinity. When a person looks at a close object by
a convergence–accommodation reflex:
   the eyes converge,
   the pupils become constricted,
   the eyes are focused on the object.

Convergence of eye
By the simultaneous action of the six extrinsic muscles of the eye (the four recti,
the two oblique) the eyes move to look at the object.

Constriction of the pupil
The pupils become constricted by an increase in parasympathetic activity and a
decrease in sympathetic activity.

Focusing
An image is formed on the retina by the curved outer surface of the cornea with
finer focusing performed by the lens, the shape of which can be altered by
relaxation or contraction of the ciliary muscle which controls the tension of the
suspensory ligament. For near vision the lens becomes fatter, for distant vision
slimmer.

Binocular vision. Two-eyed vision is better than one-eyed vision because:
(a) stereoscopic vision is provided because the retinal images in the two eyes are not
the same,
(b) size, shape and distance can be better judged,
(c) the blind spot defect (at the entry of the optic nerve) in one eye is overcome by
the other eye.

Rods and cones in the retina

The rods and cones are the light-sensitive cells in the retina. They lie side by side except in the yellow spot where there are cones only.

RODS

Rods are used for seeing in dim light and the dark and for peripheral vision. They can appreciate various shades of grey. *Visual purple* (rhodopsin) is a pigment in the rods which on exposure to dim light is broken down gradually and so sets off nervous impulses. In bright light visual purple is quickly 'bleached', i.e. broken down. *Dark adaptation*, i.e. the ability to see better in the dark after exposure to it (full adaptation takes about thirty minutes), is due to regeneration of visual purple, which occurs when a person's eye is no longer exposed to light.

Clinical features
Vitamin A is necessary for the regeneration of visual purple, and an inability to adapt to the dark is an early symptom of vitamin A deficiency.

CONES
Cones are used in bright light and to appreciate colours. How colour vision is achieved is not definitely known.

Colour-blindness is a sex-linked (X-linked) congenital defect, being transmitted by females to males. About 10 per cent of men have some degree of colour-blindness. It is rare in women. Complete colour-blindness is rare. Mild degrees may be apparent only in conditions of poor lighting or fog. The most common defect is 'red–green blindness' in which the person has difficulty in distinguishing between red, green and yellow. Other people have difficulty in distinguishing blue, green and yellow.

Clinical features
Visual acuity is tested with the aid of Snellen's types. The size and shape of each letter on them are such that any detail should subtend an angle of 1 minute when viewed at the distance stated for each line of type.

Tests for distance are done at a distance of 6 metres (Fig. 22.5). Vision is expressed as a fraction of the normal.

A normal eye can at 6 metres see the line 6 clearly.

If a person at that distance can see clearly only the letters twice that size, his vision is recorded as 6/12.

If a person can see clearly only the largest type (which a normal eye should be able to distinguish 60 metres away) his vision is recorded as 6/60.

Poorer vision is recorded as:
CF: can count fingers,
HM: can see hand movements,
PL: can perceive light.

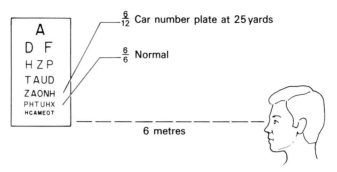

Fig. 22.5. Reading Snellen's types.

A car number plate should be readable 25 yards away.

Colour vision is detected with the use of Ishihara charts. They are composed of tiny circles in various colours and degrees of intensity, some of which are used to form numerals which the colour-blind person cannot distinguish from the background.

Blindness

The common causes of blindness in the Western world are:
    cataract: degeneration of the lens,
    glaucoma: an increase in intra-ocular pressure,
    senile degeneration of the macula of the retina.
The common causes of blindness in Africa and Central America are:
    leprosy: can cause corneal ulceration,
    trachoma: a chronic virus infection of the eye,
    onchocerciasis: invasion of eye by a filarial parasite.

23
The Ear and Hearing

EAR

Ear
    external ear
    middle ear
    inner ear

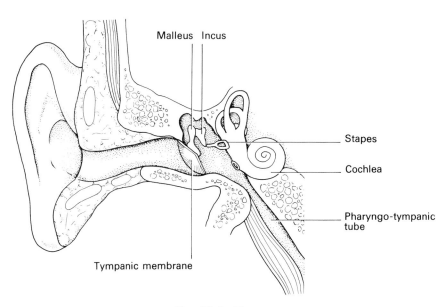

Fig. 23.1. The ear.

External ear

The external ear is composed of:
    the auricle (pinna),
    the external auditory meatus.

The *auricle* (pinna) is composed of elastic cartilage covered with skin. There is no cartilage in the lobe, which is composed only of fat and connective tissue.

The auricle can be moved slightly by three small muscles which run to it from the cranial aponeurosis and skull.

The *external auditory meatus* is the tube leading from the auricle to the tympanic membrane. The outer one-third is of cartilage, continuous with that of the auricle. The inner two-thirds are of bone. The bony part is slightly narrower than the cartilaginous part. The meatus and outer surface of the tympanic membrane are lined with skin. *Wax* is secreted by ceruminous glands in the subcutaneous tissue of the cartilaginous part.

The meatus is about 2.5 cm long and has an S-shaped bend, running first:
inwards, forwards and slightly upwards →
inwards, backwards and slightly upwards →
inwards, forwards and slightly downwards.
The meatus is oval on section at its lateral end, round at its medial end.

Middle ear (tympanic cavity)

The middle ear is a small, roughly oblong cavity in the petrous part of the temporal bone.

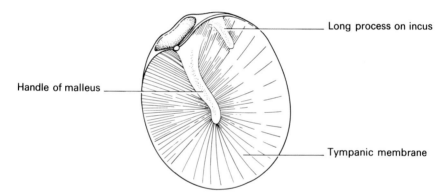

Handle of malleus

Long process on incus

Tympanic membrane

Fig. 23.2. The tympanic membrane (ear-drum).

The *tympanic membrane* (ear-drum, Fig. 23.2) occupies most of its lateral wall. It is a pearly-grey translucent membrane set obliquely across the inner end of the external auditory meatus, with its outer surface facing downwards, forwards and outwards. It is composed of connective tissue, covered on the outer surface by epithelium continuous with that of the external auditory meatus and on the inner side with epithelium continuous with that over the rest of the middle ear. Most of it is tightly stretched. The upper segment (pars flaccida, Shrapnell's membrane) is slightly flaccid.

The *ossicles* are three little bones which occupy much of the cavity, stretching across it from the tympanic membrane on the lateral wall to a little window in the medial wall.

> *Ossicles*
> malleus
> incus
> stapes

The *malleus* (hammer) has a head, a long process (handle) which is attached to the inner surface of the tympanic membrane, and a short process.

The *incus* (anvil) articulates with the head of the malleus above and by one of its two processes with the stapes, thus linking malleus with stapes.

The *stapes* (stirrup) has a small head for articulation with the incus, two processes, and a footpiece which fits into the oval window in the medial wall of the middle ear.

The ossicles are connected by synovial joints at their places of contact and are acted upon by two small muscles, the tensor tympani (inserted into the malleus) and the stapedius (inserted into the stapes).

The *epitympanic recess* is the part of the middle ear above the tympanic membrane. The head of the malleus sticks up into it.

The *mastoid opening* on the posterior wall of the middle ear leads into the mastoid antrum and other air cells in the mastoid process of the temporal bone.

The *pharyngo-tympanic tube* (*Eustachian tube*) is a bony and cartilaginous tube connecting the nasopharynx with the middle ear and allowing air to pass from one into the other. It opens into the anterior wall of the middle ear.

Internal ear

The *internal ear* (Fig. 23.3) is situated in the petrous portion of the temporal bone. It is extremely complicated. It consists of two organs:

the organ of hearing,
the organ of balance.

> *Labyrinth*
> vestibule
> cochlea
> 3 semicircular canals

The *bony labyrinth* of the internal ear is a series of interconnected cavities. The *membranous labyrinth* is a closed sac within the bony labyrinth and of approximately the same shape. The *perilymph* is a clear fluid occupying the space between bony and membranous labyrinths. The *endolymph* is a fluid lying within the membranous labyrinth.

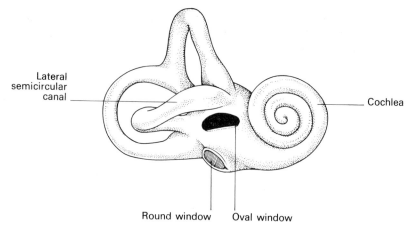

Fig. 23.3. The inner ear.

The *vestibule* is a small chamber which communicates:
anteriorly: with the cochlea
laterally: with the middle ear by two openings: (i) an oval opening closed by the footpiece of the stapes, (ii) a round opening closed by membrane.
posteriorly: with the semicircular canals.

The *cochlea* is curled like the shell of a snail. It is hollow, with the cochlear canal curling round a central pillar for $2\frac{3}{4}$ turns. Inside the bony cochlea a membranous tube runs from the bottom to the apex and then down again. The tube going up begins at the oval window and is called the scala vestibuli. The tube going down is called the scala tympani and ends at the round window.

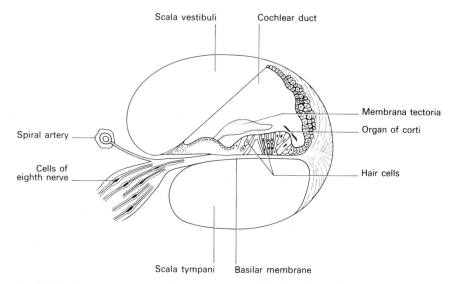

Fig. 23.4. The inner ear—cochlea. The organ of Corti is the auditory receptor organ.

The scala media is an endolymph-containing tube situated between the other two. The *organ of Corti* (Fig. 23.4) is a complicated structure which runs spirally up the cochlea, being supported all the way by the central pillar to which is attached a basilar membrane. The hair cells of the organ of Corti, numbering about 15 000, project from the basilar membrane into the scala media.

The *semicircular canals* are set at right angles to one another. They are:
a superior canal,
a lateral canal,
a posterior canal.
They contain endolymph and open into the posterior wall of the vestibule. Nerve endings of the vestibular branch of the eighth cranial nerve are connected with hair cells projecting into the endolymph.

The *saccule* and the *utricle* are parts of the membranous vestibule. They contain cells with hairs sticking into a jelly-like substance containing a number of *otoliths*, tiny crystals of calcium carbonate.

THE EIGHTH CRANIAL (AUDITORY) NERVE

The eighth cranial nerve is the nerve of the internal ear. It is composed of:
a *cochlear part* whose fibres begin around the hair cells of the cochlea and transmit impulses of hearing,
a *vestibular part* whose fibres begin in the cells of the semicircular canals and the vestibule and transmit sensations of balance.

Clinical features
Wax, the normal secretion of the glands in the external auditory meatus, can be secreted in excess and become impacted, causing irritation and deafness. *Foreign bodies* such as pips and beads can get stuck in the meatus.

Acute otitis media is an acute infection of the middle ear, which is most commonly infected from the nasopharynx via the pharyngotympanic (Eustachian) tube. It causes earache and deafness. The tympanic membrane is liable to bulge and rupture under the increased pressure. A *perforation* of the tympanic membrane is the hole thus produced. *Chronic otitis media* is a chronic infection of the middle ear which may follow an acute otitis media. Infection can spread into the mastoid process or through the skull into the meninges and brain.

HEARING

Hearing is the ability to detect certain pressure vibrations in the air and to interpret them as sound. The ear converts the energy of the pressure waves into nervous impulses, and the cerebral cortex converts these impulses into sounds.

Sounds possess frequency, amplitude and wave-form.

The *frequency* of a sound-wave is the rate at which air-waves oscillate per unit time.

The human ear can pick up frequencies varying from about 20 to 16 000 Hertz (Hz). A Hertz is a cycle per second. Sounds of low frequency have low 'pitch'. Sounds of high frequency have a high 'pitch'.

The human voice ranges from about 65 Hz to just over 1000 Hz. The human frequency mechanism is most sensitive to sounds with frequency of about 1000 Hz.

The *amplitude* is a measure of the energy or intensity of the pressure fluctuations. Sound waves of different amplitude are interpreted as differences in loudness.

Measurement of sounds
sounds are measured in decibels (dB)
    whisper   about 20 dB
    quiet conversation   about 50 dB
    noisy factory   about 100 dB
sounds above 120 dB cause pain and prolonged exposure can damage the ear and cause deafness

---

Wave-forms are sound waves interpreted as having different qualities, such as pure tone, musical sounds or (when there is no regularly recurring pattern) noise.

Sound transmission in the ear

EXTERNAL EAR

Sound waves are picked up by the auricle and transmitted down the external auditory meatus.

TYMPANIC MEMBRANE

Sound waves cause the tympanic membrane to vibrate (Fig. 23.5). The membrane is an elastic membrane, which has no natural frequency of its own but takes up the characteristics of the vibrations applied to it. It can vibrate readily because the pressure on both its sides is atmospheric. The pharyngeal end of the Eustachian tube is opened by swallowing, sneezing and yawning, and so long as the tube is patent the middle ear is kept filled with air at atmospheric pressure. The membrane will not vibrate so well if the tube is blocked and the pressure on the two sides unequal. The amplitude of the vibration of the membrane is proportional to the intensity of the sound. The membrane is highly damped, i.e. it stops vibrating as soon as the sound stops.

OSSICLES

The vibrations of the tympanic membrane are picked up by the malleus, which is attached to its inner surface and transmitted through the incus to the stapes.

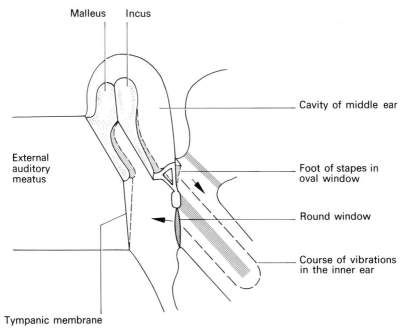

Fig. 23.5. Diagram to show the course taken by auditory vibrations. The cochlea is represented by a simple tube to show the course of vibrations in the inner ear.

The footpiece of the stapes transmits the vibrations through the oval window into which it is set. The area of the tympanic membrane is 15–20 times larger than that of the oval window, and the force of the vibration at the window is greater than at the tympanic membrane, although there is some loss of energy due to the inertia of the ossicles.

The stapedius and tensor tympani muscles contract reflexly in response to loud noises, and by contracting pull on the ossicles, making the ossicular system more rigid and so protecting the inner ear.

COCHLEA

Vibrations at the oval window set up pressure waves in the perilymph of the inner ear. Waves travel up the perilymph in the scala vestibuli and down the perilymph in the scala tympani. When a wave reaches the round window at the bottom, the membrane closing the window gives a little, bulging into the middle ear. If it did not give, waves could not pass along the cochlea.

ORGAN OF CORTI

The manner in which the organ of Corti responds to vibrations is not known with certainty. A movement of the basilar membrane, produced by waves passing up and down the cochlea, appears to pull on the hair cells and excite them to transmit impulses into the nerve fibres of the cochlear nerve situated around the bases of the hair cells.

According to the 'travelling wave theory', waves produced by high frequency sound travel only a little way in the cochlea before being damped out, and those of low frequency travel up to the apex of the cochlea. The discrimination by the ear between sounds of differing frequencies seems to be the result of differing patterns of vibration set up in the basilar membrane by differing frequencies.

CENTRAL CONNECTIONS

The cochlear part of the auditory nerve transmits sensations to the brain, where they are interpreted in the auditory part of the temporal lobe. Each ear has connections with both temporal lobes, and especially with the opposite one.

Clinical features
Deafness can be due to:
(a) wax in the external auditory meatus,
(b) otitis media,
(c) otosclerosis, a condition in which new bone forming round the footpiece of the stapes prevents it from moving,
(d) injuries of the tympanic membrane,
(e) injury, disease or degeneration of the auditory nerve.

Tinnitus is a ringing, buzzing, hissing or pulsating noise in the ear. It can be a symptom of any abnormal condition of the ear.

BALANCE

The semicircular canals, the saccule and the utricle are concerned with balance and the position of the head on the shoulders.

The *semicircular canals* are involved in rotatory movements of the head. These movements produce movements of the endolymph in the semicircular canals, which stimulate the hair cells. The *otoliths* of saccule and utricle are moved by changes in the position of the head and so provoke movements of the hair cells there.

Stimuli are transmitted along the nerve fibres of the vestibular branch of the eighth (auditory) cranial nerve to the midbrain, medulla oblongata, cerebellum and spinal cord. These stimuli initiate reflex alterations in the muscles of the neck, eye, trunk and limbs in order that balance and posture are maintained and the eyes can be fixed on a moving object.

Clinical features
Giddiness (vertigo) is a symptom of disorder or disease of the organs of balance. In *Ménière's disease* an increase of endolymphatic pressure causes severe attacks

of giddiness, often associated with nausea and vomiting and sometimes with deafness and tinnitus. *Infections of the labyrinth, fractures of the petrous part of the temporal bone, atheroma of the vertebral artery causing a reduction of blood supply to the internal ear, and new growths* will cause vertigo. *Wax* in the external auditory meatus pressing on the tympanic membrane and blocking of the pharyngotympanic tube can cause slight deafness. *Motion sickness* (car, train, aeroplane, sea sickness) is the result of labyrinthine disturbance by repeated movement of the endolymph, often associated with a strong emotional factor so that a susceptible person may be sick or have giddiness with the anticipation of the movement. Various drugs can prevent motion sickness, but how they work is not known. Some drugs (especially streptomycin) can cause labyrinthine degeneration.

24

The Defences of the Body

The defences of the body are against:
injury
infection and invasion by foreign cells
new growths

INJURY

The tissues and organs of the body may be damaged by cuts, wounds and burns.

The body attempts to restore damaged tissues and organs to normal structure and function by:

(a) removing cells so damaged that they cannot function again,

(b) replacing them with new cells.

The removal of damaged cells is achieved by macrophages and other phagocytic cells, i.e. those cells whose function it is to absorb within themselves and then destroy dead or dying cells and tiny particles of foreign matter.

The replacement of old cells by new cells depends upon whether the specialized cells of the tissue can divide into new functioning cells or not. Cells are divided into:

labile cells which divide freely, e.g. those of the liver,

stable cells which divide only after a strong stimulus,

permanent cells which never divide, e.g. those of the nervous system.

If a damaged tissue is composed of labile cells, a perfect or near-perfect replacement with normally functioning cells is possible. When this is not possible as with stable or permanent cells, fibrous tissue is formed which replaces the damaged tissue but can never take over the specialized functions performed previously by the damaged cells.

Essentials for rapid sound healing
adequate protein and vitamin C intake by
patient
normally functioning thyroid gland
adequate blood supply to damaged tissue
no excessive haemorrhage into damaged tissue
no infection of damaged tissue

Healing in skin

The healing processes are seen in their simplest form in the healing of a small cut of the skin.

The following processes take place:

1. Plasma, plus some red blood cells, pours out of the adjacent capillaries, clots and plugs the gap by binding the two sides of it together.

2. Solid buds of cytoplasm grow out of the capillaries on both sides of the cut and fuse in the middle.

3. These buds become canalized, and when this has happened blood flows through the new channels.

4. Macrophages migrate into the clot to absorb the debris of cells and any small foreign particles.

5. Fibroblasts grow within the clot at right angles to the original cut and manufacture collagen.

6. The epithelial cells of the skin on both sides of the cut grow towards one another and fuse to cover the cut on the outside.

7. The collagen contracts to form a thin scar.

8. With the disappearance now of the macrophages and of any excess of blood vessels, the tissues become normal.

INFECTIONS AND INVASION BY FOREIGN CELLS

Immunity is the property of the body to defend itself against infection and attempts to introduce into it cells from other people or animals.

The body is under constant attack by micro-organisms. Harmful organisms can invade the body and if they are not checked, they multiply in it and cause illness or death.

Resistance to infection by micro-organisms is achieved by:

(a) non-specific factors

(b) specific factors, i.e. the production of immunoglobulins.

Non-specific factors in resistance

(a) Lactic acid and fatty acids in sweat and sebaceous secretion keep the pH of the skin low. Most micro-organisms cannot survive in such conditions and die.

(b) Micro-organisms which enter the respiratory tract are wafted away from the lungs by the movements of the cilia that stick out of the epithelial cells lining the tract.

(c) Tears and saliva contain lysozyme, an antibacterial enzyme.

(d) Mucus secretions on the surfaces of cells prevent the invasion of the cells by viruses.

(e) Granulocytes and macrophages can recognize micro-organisms as enemies and engulf and destroy them.

(f) The invasion of a tissue by micro-organisms stimulates an inflammatory response:

  (i) the walls of the capillaries become more permeable,
  (ii) granulocytes and monocytes move out of them into the invaded tissues,
  (iii) plasma with various bactericidal factors in it passes into the tissues.

Interferon is an anti-viral agent which is not specific against any one virus, but is produced by cells when they are invaded by viruses. It appears to be important in recovery from viral infections. It may have anti-cancer properties.

Specific factors in resistance

THE DEVELOPMENT OF IMMUNITY

The immunity of the body to infection is due to the development in it of anti-infective factors. Specific factors develop after an attack by a particular kind of micro-organism.

  The reaction is: antigen→antibody

An *antigen* is a protein within a micro-organism which produces in the body an *antibody*. The presence of an antibody in sufficient amounts will prevent or modify any subsequent attack by that micro-organism.

  The reaction is a specific one: it is limited to one micro-organism and does not give protection against an unrelated micro-organism.

IMMUNOGLOBULINS

The antibodies are immunoglobulins, which belong to the gamma globulin class. They are produced in:

  lymph tissue in the spleen   lymph nodes
  lymph tissue in the wall of the intestine bone marrow
  There are five major types. (Ig = immunoglobulin)

IgG

IgG is the most common Ig of body fluids. It is present in blood and tissue-fluids in about equal proportions. It circulates continuously through the capillary walls into fluid, into lymph and back to the blood via the thoracic duct. It combats micro-organisms and their toxins (poisonous secretions). It promotes phagocytosis by polymorphs and macrophages. It can pass through the placenta to protect the fetus, being the only Ig capable of doing this.

IgA

Much of IgA is produced in the wall of the small intestine. It passes into seromucous secretions and is found in tears, saliva, nasal secretions, colostrum and milk, secretions from the lungs and alimentary tract. Its main task is the defence of exposed surfaces against micro-organisms.

IgM

IgM is a first-line defence against micro-organisms circulating and attempting to multiply in the blood. It appears early in an infection and stays mostly in the blood.

IgD

Antibody function of IgD is uncertain.

IgE

The main antibody function of IgE is uncertain. It is responsible for the symptoms of extrinsic asthma and hay fever which occur when a patient is exposed to an allergen he is sensitive to, e.g. grass pollen in hay fever. It becomes raised when the body is infected with certain parasites.

THYMUS

The thymus (Fig. 24.1) is a soft multi-lobed organ. It is relatively large at birth, increases in size up to about 16 years, and then becomes smaller. It lies in front of the neck and upper part of the thorax, and at its largest can extend from the lower border of the thyroid cartilage to the upper border of the pericardium.

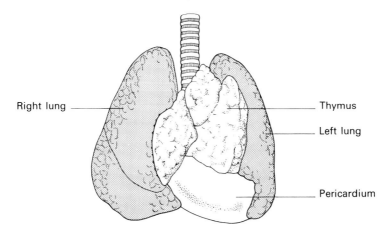

Right lung

Thymus

Left lung

Pericardium

Fig. 24.1. The thymus in early life.

Its lobes are connected by connective tissue. In childhood it is composed of a cortex of tightly packed lymphocytes and a medulla of loose tissue with thymic (Hassall's) corpuscles, which are groups of concentrically arranged cells. With

---

Functions of thymus
source of all lymphocytes before birth
source of long-lived lymphocytes after birth
'processes' T lymphocytes

age the difference between cortex and medulla disappears, the gland then consisting of connective tissue and a few lymphocytes and thymic corpuscles.

LYMPHOCYTES

Lymphocytes (Fig. 24.2) vary in diameter from 5–15 μm. They are round cells with a round or slightly indented nucleus, which occupies most of the cell, the

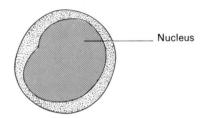

Nucleus

Fig. 24.2. A lymphocyte.

cytoplasm being no more than a narrow rim around it. Although the functions of different lymphocytes may be different they all look the same except for the variation in size. They are found in lymph tissue (where they are the most numerous of cells there), lymph and blood.

Lymphocytes are essential for the development of immunity, but all their activities are not definitely known. Some lymphocytes, called natural killer cells, can identify and destroy cancer cells, but do not attack normal cells. Lymphokines are substances which are produced by lymphocytes and are thought to coordinate and amplify immune responses.

There appear to be two types:

Short-lived lymphocytes, which are produced in bone marrow and live for only a few days.

Long-lived lymphocytes, which are produced in the thymus and other lymph tissue and live for months, possibly for years.

The long-lived lymphocytes are divided into:

T lymphocytes (T = thymus). These form about two-thirds of all lymphocytes in the blood. In order to perform their immunity functions they have to be 'processed' by the thymus and have to circulate in and out of it. They appear to be able to recognize an antigen and to start off the preliminary reaction to it.

B lymphocytes (B = bone marrow in man). These form about one-third of the lymphocytes in the blood and originate in bone marrow. They do not have to pass into the thymus to become active. They are the antibody-producing lymphocytes and carry upon their outer surface the immunoglobulins which they have formed.

Many of the lymphocytes available in lymph tissue throughout the body at any one time have not produced antibodies. They form a pool of unstimulated cells, standing by to produce immunoglobulins in response to any new infection that might attack the body.

MACROPHAGES

Macrophages are large cells, with phagocytic properties, i.e. they are capable of absorbing and destroying micro-organisms. They occur as:

free macrophages: cells capable of moving freely about. The monocytes of the blood and the histocytes found in interstitial tissue are free macrophages.

fixed macrophages: not so fixed as their name suggests, but not as mobile as the free macrophages. They are found in the interstitial tissue of the spleen and lymph nodes and in the liver, where they form part of the lining of the blood vessels.

PREVENTION OF COMMUNICABLE DISEASES

Immunity to a communicable disease may be inherited, it may be acquired naturally by having an attack of the disease, it may be produced artificially by the injection into the body of antigens in order to stimulate the production of the specific antibodies to them.

The antigen takes the form of a 'vaccine' composed of either:
(a) live organisms whose ability to produce disease has been weakened by their being grown in adverse conditions; or
(b) dead organisms which no longer have the power to produce the disease but have still the power to stimulate an appropriate antibody reaction.

Diseases preventible by vaccination	
diphtheria	whooping cough
measles	poliomyelitis
tetanus	German measles
tuberculosis	yellow fever

The amounts of antibody produced in one of these ways are not usually so great as those produced by actual infections, and the protection given is not so great nor lasts so long. Booster doses are necessary later if the antibody-content is to be maintained at any adequate level.

TRANSPLANTS

The transplants of tissues or organs from one person to another are subject to the same laws that govern immune phenomena. The body regards the transplanted cells as if they were invading micro-organisms.

Transplanted tissues and organs	
skin	heart
kidney	heart and lung together
cornea	red bone marrow
liver	blood (in a transfusion)

The following terms are used:

autograft: a graft from one part to another of the same person,

isograft: a graft between identical twins,

allograft: a graft between members of the same species, e.g. from man to man,

heterograft (xerograft): a graft between members of different species, e.g. from pig to man.

In an *autograft*, which is of skin, there should be no rejection provided that the part does not become infected and that it is not subjected to tensions which would pull it off.

In an *isograft* there should not be an immunological reaction because the tissues and organs of identical twins, having been derived from the same fertilized ovum, are immunologically identical.

In *allografts* and *heterografts* immunological reactions will inevitably occur unless the body into which the tissue or organ is to be transplanted has been prepared in such a way that there is no immunological reaction: such a preparation is neither easy nor safe.

Tissues, such as the cornea, which have no blood supply do not stimulate rejection reactions. The recipient body treats all other grafts tissues as invaders. It recognizes the new cells as 'not-self' cells. Lymphocytes immigrate into the graft in large numbers; clots of blood appear in its blood vessels; and the tissue then ceases to function and dies.

To prevent this reaction an immunosuppressive drug can be given, i.e. one which prevents the invasion of the graft by lymphocytes by stopping the development of lymphocytes. This inevitably interferes with the body's normal defence mechanism against infection.

AUTOIMMUNE DISEASES

In the diseases called autoimmune diseases it is thought that certain antibodies are produced which can attack cells of the same body, treating them as if they were invaders.

Among the diseases thought to belong to this class are:

rheumatoid arthritis   pernicious anaemia (most cases)

myasthenia gravis   systemic lupus erythematosus.

NEW GROWTHS

The defences of the body against cancer are imperfectly understood. It is possible that abnormal cells are produced not infrequently and that the body has a defence against them in the form of lymphocytes which are thought to be continually policing tissues, sniffing out any abnormal cells and destroying them. The occurrence of a cancer in a person may be the result of a breakdown of this defence mechanism and an unchecked proliferation of abnormal cells. Some human cancer cells seem to be able to produce antibodies against themselves.

25
Reproduction

Male genital system
   testes
   epididymis and other ducts
   prostate gland
   penis

The testes

The testes (Fig. 25.2) are oval bodies, each of which is suspended by the spermatic cord in its half of the scrotum.

The scrotum is the pouch in which lie the testes. Its skin is thin and pigmented. The *cremasteric muscle* is a thin sheet of muscle in the scrotum which by contracting lifts up the testes.

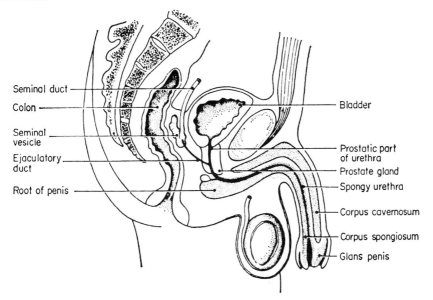

Fig. 25.1. The male genital system.

277

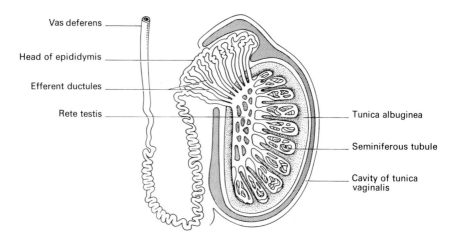

Fig. 25.2. The testis and epididymis.

The *tunica albuginea* is the fibrous capsule of the testis. The *tunica vaginalis* is a double covering, with a potential space between the two layers, which surround the testis except posteriorly.

STRUCTURE

(a) Each testis is divided by septa into 200–300 lobules. Each lobule contains 1–3 seminiferous tubules.
(b) The seminiferous tubules are coiled tubules, up to 70 cm long.
(c) Spermatozoa are produced after puberty by cells of the seminiferous tubules.
(d) At the back of the testis the seminiferous tubules open into about 12 efferent ducts, which pierce the tunica albuginea and form the head of the epididymis.
(e) Interstitial cells, which produce the male hormone, lie between the tubules.

BLOOD SUPPLY

Each testicular artery arises from the abdominal aorta just below the origin of the renal arteries, passes down in front of the posterior abdominal wall, and reaches the testis and epididymis by passing down the spermatic cord.

VENOUS DRAINAGE

Blood from the testis passes into the pampiniform plexus of veins in the spermatic cord. This plexus drains into a single vein, which passes up the back of the abdomen. The right testicular vein enters the inferior vena cava. The left enters the left renal vein.

LYMPH DRAINAGE

Lymph vessels pass up in the spermatic cord to enter aortic lymph nodes.

NERVE SUPPLY

Sympathetic nerves pass to the testis on the testicular artery.

Clinical feature

A *hydrocele* is a collection of fluid within the sac of the tunica vaginalis.

Spermatozoa are formed after puberty in the cells which form the walls of the seminiferous tubules and pass along the tubules into the efferent ducts and so into the epididymis.

Testosterone, the male hormone, is produced in the interstitial cells of the testis from puberty onwards (see p. 217).

The epididymis and other ducts

These are a series of tubes through which the spermatozoa pass.

Epididymis: the collecting organ attached to the back of the testis. It has a head (composed of the efferent tubules coming from the testis), a body, and a tail (composed of the single tube into which the ducts run).

Vas deferens: a thick-walled tube. It begins at the lower end of the epididymis as a continuation of its tube, passes upwards behind the epididymis, runs up the inguinal canal, enters the pelvis, and passes to the back of the bladder, where it lies medial to the seminal vesicle of its own side.

Seminal vesicle: a vessel formed of coiled sacculated tubes, lying, one on each side, at the back of the bladder.

Ejaculatory duct: a common duct for the vas deferens and the seminal vesicle. It passes through the prostate gland to open into the prostatic part of the urethra.

Spermatic cord
   vas deferens
   testicular artery
   pampiniform plexus of veins
   lymph vessels of the testis
   nerves

Prostate gland

The prostate gland is about the shape and size of a horse chestnut. It surrounds the first part of the urethra.

   It lies:

below the bladder,

behind the symphysis pubis,

in front of the rectum.

It is traversed by the urethra and the ejaculatory ducts.

It is composed of a number of tubular glands and fibromuscular tissue, the whole enclosed within a capsule.

Clinical features

The gland can be felt as a firm smooth object by rectal examination. *Enlargement of the gland* is common after the age of fifty and can cause obstruction to micturition by constricting the urethra. *Cancer* of the gland causes a similar obstruction and can spread to other organs and tissues.

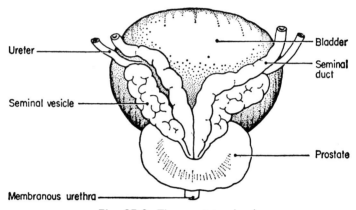

Fig. 25.3. The prostate gland.

Penis

The penis is composed of three cylindrical bodies—the right and left corpora cavernosa and a central corpus spongiosum. It is attached posteriorly to the sides of the pubic bone and to the perineum.

The glans penis is a knob of tissue into which the corpus spongiosum is enlarged at the end of the penis. It is enclosed in the prepuce.

The urethra enters the corpus spongiosum posteriorly, runs its entire length, and opens at the external urethral opening on the tip of the glans penis.

COMPOSITION

The corpora cavernosa and corpus spongiosum are composed of a sponge-like tissue, formed of connective tissue and smooth muscle. This tissue encloses dilatable vascular spaces.

Clinical feature

Circumcision is removal of the prepuce.

Spermatozoa are formed in the seminiferous tubules and pass into the epididymis and along the vas deferens. The seminal vesicles do not store the spermatozoa, but produce some of the fluid in which the spermatozoa are carried along. The prostate gland produces a similar fluid.

Under sexual excitement impulses pass along the parasympathetic nerves to the arterioles of the penis. The arterioles dilate and the vascular spaces of the penis become engorged with blood and the penis enlarges to become stiff and erect. Sexual intercourse is then possible. At ejaculation the semen, composed of sperm in the fluid produced by the seminal vesicles and prostate gland, is discharged through the urethra.

FEMALE GENITAL SYSTEM

> *Female genital system*
>   right and left ovaries
>   right and left uterine tubes
>   uterus
>   vagina
>   external genitalia
> *Accessory organs*: the breasts

Ovary

There are two ovaries, right and left. Each is about 4 cm long, 1.5 cm wide, and 1 cm thick.

The ovary is attached by a fold of peritoneum to the back of the broad ligament, a double fold of peritoneum stretching from the side of the uterus to the lateral wall of the pelvis. The uterine tube arches over the ovary and ends on it. In women who have not borne children, the ovary usually lies in the ovarian fossa, a shallow peritoneal depression in the lateral wall of the pelvis. After childbirth its position is variable.

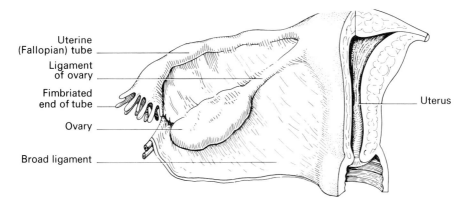

Fig. 25.4. The female sex organs within the pelvis.

STRUCTURE

Before puberty
The ovary consists of an external layer of germinal cells and primary undeveloped ovarian follicles which surround a central medulla of fibrous tissue, unstriated muscle fibres and blood vessels.

From puberty to the menopause
The ovarian follicles are in various stages of development. A *Graafian follicle* is a mature follicle, composed of an ovary and fluid within a layer of cells. A *corpus luteum* is a follicle from which the ovum has been discharged. If the woman becomes pregnant, the corpus luteum enlarges to a diameter of about 5 cm; if she does not become pregnant, it degenerates. With the discharge of more and more ova, the surface of the ovary becomes puckered with the scars of old corpora lutea.

After the menopause
Follicles are no longer formed, and the ovary shrinks.

Uterine (fallopian) tube

There are two uterine tubes, right and left. Each is about 20 cm long and extends from the lateral angle of the uterus to the ovary, lying in the free upper margin of the broad ligament. At the ovarian end the tube opens by a small hole into the peritoneal cavity. This hole is surrounded by fimbriae, frond-like moving processes.

STRUCTURE

The tube is composed of an inner layer of ciliated cells, a middle layer of muscle, and an outer layer of peritoneum.

When an ovarian follicle is ripe, the fimbriae close round it so that on its discharge from the ovary the ovum passes straight into the tube. It is then moved along the tube by contractions of the muscle in the wall and by movements of the cilia and is eventually passed into the uterus.

Clinical features
An *ectopic pregnancy* is a pregnancy in which the fertilized ovum remains in the tube instead of moving out of it into the uterus. Causes of this can be: (a) an infection that has damaged the cilia and stopped them from working, (b) a partial obstruction or kinking of the tube by adhesions. At about the 6th week the enlarging ovum grows too big for the tube and bursts through the wall. This is called a *ruptured ectopic pregnancy*. It causes a severe haemorrhage which can be fatal if the damaged tube is not removed.

Uterus

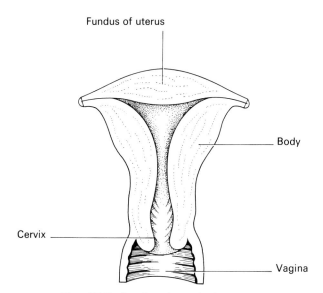

Fundus of uterus

Body

Cervix

Vagina

Fig. 25.5. Section through the uterus.

The uterus (Fig. 25.5) is a thick-walled muscular pear-shaped organ about 7.5 cm long. It lies in the pelvis between the bladder in front and the rectum behind (Fig. 25.6). It consists of:

the *fundus*: the rounded upper end, above the line of attachment of the uterine tubes.

the *body*: forms about two-thirds of the organ. It is continuous with the cervix at an angle, so that the whole uterus bends forwards (anteversion). It has an antero-inferior surface, which lies on the bladder, and a posterosuperior surface in contact with coils of small intestine.

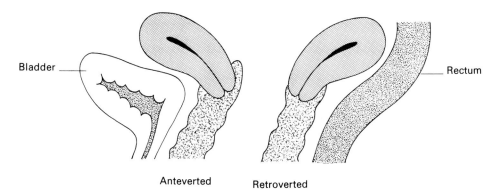

Bladder

Rectum

Anteverted   Retroverted

Fig. 25.6. The normal anteverted uterus and a retroverted uterus.

the *cervix*: the cylindrical lower one-third. The lower end of it sticks backwards and downwards into the vagina.

The uterus is hollow, but its walls are closely approximated. In the non-pregnant uterus the space in the body is only a triangular chink. The cervix has a cervical canal between the *internal os* (where the cervix opens into the body) and the *external os* (where it opens into the vagina).

Peritoneal attachments of the uterus

The *broad ligament* on each side runs from the lateral border of the uterus to the lateral wall of the pelvis, forming a partition across the pelvis. The uterine blood vessels, lymph vessels and nerves run between its two layers. The ovarian tube lies in its upper border.

The *round ligament* is a narrow fibrous band which runs from the lateral angle of the uterus, through the broad ligament, and then down the inguinal canal to end in the labium major.

The *recto-uterine pouch* (of Douglas) is formed by the reflection of peritoneum from the rectum on to the uterus.

BLOOD SUPPLY

The uterine artery on each side is a branch of the internal iliac artery. It runs between the layers of the broad ligament. It anastomoses with the ovarian artery, which supplies the fundus. It is tortuous to allow for stretching during pregnancy.

VENOUS DRAINAGE

Into the internal iliac vein.

LYMPH DRAINAGE

(a) From the fundus lymph vessels run with the round ligament and end in inguinal lymph nodes; (b) from the body lymph vessels run in the broad ligament and reach the aortic lymph nodes; (c) from the cervix lymph vessels run to external iliac, internal iliac and sacral lymph nodes.

STRUCTURE

Endometrium: the inner layer of columnar epithelial cells and glands; shows changes at various stages of the menstrual cycle.

Myometrium: the middle layer of plain muscle fibres, arranged in broad interlacing bundles.

Peritoneum: covers the exterior, except for the intravaginal part of the cervix.

Functions of uterus
1. The endometrium of the uterus undergoes changes in order that it may be prepared to receive a fertilized ovum.
2. The endometrium receives the fertilized ovum and the uterus develops and enlarges in order to retain and nourish the developing embryo and fetus.
3. During childbirth contractions of the myometrium expel the fetus.

Menstruation

Menstruation begins at puberty about the age of 11–12 years and recurs at intervals of about 28 days until the menopause about the age of 45–50 years. The regularity of the cycle is due to the regular occurrence of ovulation. This is controlled by two anterior pituitary hormones:
follicle stimulating hormone (FSH)
luteinizing hormone (LH).

Endometrial phases
proliferative (pre-ovular) phase
secretory (post-ovular) phase
menstrual phase

PROLIFERATIVE (PRE-OVULAR) PHASE

Duration: about 14 days.
Hormonal activity: oestrogen is secreted by the ovarian follicle under the influence of FSH.

The phase lasts from the end of menstruation to the discharge of the ovum from the ovary. There is a rapid growth of endometrium, the whole of the interior of the uterus becoming lined with a layer within two days. The layer is at first thin and composed of cuboidal cells; but as the phase proceeds the cells become columnar, the glands in the endometrium lengthen, and the whole endometrium becomes thicker.

Mittelschmerz (German for mid-pain) is a pain (like a stitch) in the lower abdomen felt by many women at the end of this phase. It occurs just before the ovarian follicle ruptures; its cause is uncertain: it has been attributed to a hormonally produced contraction of smooth muscle fibres around the follicle. It can help to identify the most fertile day of the mid-cycle and so help natural family planning or artificial insemination.

SECRETORY (POST-OVULAR) PHASE

Duration: about 13 days.
Hormonal activity. Progesterone and a little oestrogen are produced by the

corpus luteum in the ovary. If the ovum is not fertilized, the corpus luteum shrinks, and from about the 22nd day the amount of progesterone secreted starts to fall.

Endometrial development continues. It becomes more vascular. The glands in it become distended with secretion.

MENSTRUAL PHASE

Duration: usually 4–5 days.
Hormonal activity: the amount of progesterone in the blood continues to fall.

The endometrium degenerates, glandular secretions are discharged, and the unsupported capillaries break down and bleed.

Composition of menstrual fluid
  blood
  secretion from endometrial glands
  degenerated endometrium
  secretion from cervical glands
  vaginal fluid
The blood does not clot because it becomes
defibrinated by contact with endometrial tissue)

Clinical features
The cervix can be felt by a finger introduced into the vagina or rectum. The whole uterus can be felt bimanually with a finger in the vagina or rectum and the other hand on the anterior abdominal wall.

During pregnancy the uterus:
(a) appears above the symphysis pubis at the 12th week,
(b) reaches the umbilicus at the 24th week,
(c) reaches the xiphisternum at the 36th week.

A *fibroid* (fibroymyoma) is a non-malignant tumour of fibrous and muscular tissue growing in the myometrium of the uterus; it is likely to cause excessive menstrual bleeding, bleeding between periods, and mechanical obstruction to structures in the pelvis and abdomen. *Cancer* of the body and of the cervix are common forms of cancer.

Vagina

The vagina is a tube which extends from the cervix of the uterus to the vestibule of the vulva. It is shorter in front than behind, its anterior wall being about 7.5 cm long and its posterior wall 9.0 cm long. The cervix of the uterus projects into the upper part of the vagina. The *anterior*, *lateral* and *posterior fornices* are those part of the vagina that are in front of, at the side of, and behind the cervix respectively. The *hymen* is a thin mucosal fold at the opening of the vagina.

> *Relations of vagina*
> *anterior:*   bladder and urethra
> *posterior:*   recto-uterine pouch (of Douglas),
>   of peritoneum
>   rectum
>   anal canal
> *lateral:*   pelvic fascia
>   ureter (close to the lateral fornix)
>   levator ani muscle

STRUCTURE

(a) Mucous membrane, of squamous epithelium,
(b) Muscular coat, of plain muscle,
(c) Fibrous tissue, blending with the fascia of the pelvis.

The vagina is kept moist by secretion from the cervix of the uterus and by a transudation of tissue-fluid through the vaginal wall. The secretion is made acid by bacterial action.

Clinical features
The acidity of the vagina fluid helps to make the vagina resistant to infection. Rarely the hymen extends across the opening of the vagina, preventing the escape of menstrual fluid.

Vulva

The *vulva* is the name given to the female external genitalia. It consists of:

labia majora: two large hair-bearing folds which extend from the mons pubis (the pad of fat in front of the symphysis pubis) to the perineum in the midline behind.

labia minora: two thin lips of pigmented soft skin lying within the labia majora, dividing in front to enclose the clitoris and meeting behind at the fourchette, a short transverse fold.

clitoris: a small erectile organ in the midline in front; the equivalent of the penis of the male.

vestibule: the area enclosed by the labia minora and containing: (i) the opening of the urethra, just behind the clitoris, and (ii) the opening of the vagina.

Bartholin's glands: a pair of oval mucus-secreting glands lying deep to the posterior parts of the labia majora and opening by a duct at the side of the labia minora.

Clinical features
A *Bartholin's cyst* is a cyst of the greater vestibular gland due to obstruction of the duct. A *Bartholin's abscess* is an abscess of the gland.

BREAST (MAMMA)

The female breast extends from the 2nd to the 6th ribs and from the sides of the sternum to the mid-axillary line. It lies on deep fascia, which separates it from the pectoralis major and other muscles. The *areola* is the pigmented area around the nipple. The *areolar glands* (of Montgomery) are large sebaceous glands in the areola.

In a man the breast is rudimentary.

STRUCTURE

The breast is composed of 15–20 lobules of glandular tissue, embedded in fat. The lobules are separated from one another by fibrous tissue. They converge towards the nipple, and each opens into a duct which opens on the nipple.

BLOOD SUPPLY

By arteries from the axillary artery and by perforating branches of the internal mammary artery, which runs down inside the chest and behind the costal cartilages and sends off branches through the intercostal spaces.

VENOUS DRAINAGE

Into large veins on the surface of the breast and thence into veins corresponding with the arteries.

LYMPH DRAINAGE

Into (a) axillary lymph nodes and thence to cervical lymph nodes and (b) through the intercostal spaces to lymph nodes on the line of the internal mammary artery.

In girls the breasts develop at puberty under the influence of two hormones, oestrogen and progesterone. In pregnancy the further development is stimulated by raised oestrogen and progesterone levels in the blood. After birth *colostrum*, a yellow fluid containing fat and protein, is secreted first. The secretion of *milk* does not start until a few days later; it is the result of stimulation by prolactin of the anterior pituitary. An excessively high level of oestrogen inhibits the secretion of milk, and artificial oestrogens are given to prevent or stop lactation. Oxytocin (produced by the posterior pituitary lobe) expels milk from the mammary ducts.

Clinical features
Accessory nipples are present in some people as little pigmented functionless projections on a line between the middle of the clavicle and the symphysis pubis.

Mastitis is inflammation of the breast. *Cancer of the breast* is common: it is likely to produce a hard lump, an abnormal position of the nipple, a blood-stained discharge from the nipple, invasion of the skin. It can spread to other parts of the body, especially along the lymph vessels draining the breast. *Mastectomy* is surgical removal of the breast.

26
Heredity

Heredity is the genetic endowment transmitted by a person to his or her children.

Chromosomes

Chromosomes are organized structures present in the nucleus of every cell.

Chromosomes can be clearly demonstrated only at *mitosis*, the stage in which a cell is dividing into two daughter cells. At that time they can be stained and identified as separate objects (chromos = colour). When a cell is not dividing, the chromosomes are a tangle of thin threads which cannot be separately distinguished; it is when they are in this form that the chromosomes are engaged in their chemical activities.

In *mitosis*:

(a) a cell starts to divide into daughter cells (Fig. 26.1).

(b) the chromosomes have become shorter, rod-like structures and each divides into two longitudinally. The *centromere* of a chromosome is the last spot on a chromosome to divide.

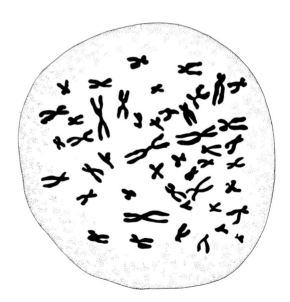

Fig. 26.1. Chromosomes as seen within the nucleus of a cell at a time of division.

(c) one half of each chromosome is attracted into each of the daughter cells.

(d) the two cells separate, each being identical with the cell from which they have come and with a full complement of chromosomes.

(e) the chromosomes revert to their former thread-like state.

Chromosomes are composed of:

DNA (deoxyribonucleic acid),

RNA (ribonucleic acid),

several kinds of protein.

DNA and RNA are members of a group of biochemical compounds called nucleic acids or polynucleotides. The DNA of a cell contains in a sort of code all the information for the synthesis of all proteins ever to be made by that cell. Every cell has this code and uses it when it needs to. DNA can both store and reduplicate information. Its structure is such that it can divide into two, and at each cell division each parent cell hands on identical copies of its own DNA to the daughter cells.

The function of RNA is that of a carrier between DNA and protein. It passes along in appropriate sequence the various pieces of information a cell requires to carry out its various activities.

Genes

Genes are the elements by means of which inherited factors are transmitted.

Genes are arranged in lines along the chromosomes and the chromosomes of each cell are thought to have 4 600 000 genes. Each gene is thought to be responsible for one particular type of enzyme reaction in the cells.

An enzyme is a biological catalyst which accelerates a specific biochemical change. The enzyme itself remains unchanged by the reaction.

By controlling all enzyme reactions the genes control all chemical changes in the body. All genes are not in action at the same time; most of them at any one time are in an inactive state, becoming active only when there is a need for a particular change to take place in a cell.

Chromosomes in man

In a human cell there are 46 chromosomes. Of these:

44 are autosomes (non-sex) chromosomes,

2 are sex chromosomes.

In a *female* the two sex chromosomes are identical and are called XX.

In a *male* there is one X chromosome (identical with those in a female) and one Y chromosome. A Y chromosome is much smaller than an X chromosome.

To obtain human chromosomes blood or skin is used: skin is more difficult to use but necessary for certain investigations. White blood cells will, if cultured in a certain medium, show mitosis. Complete division is stopped by adding certain chemicals. Specimens are examined under a microscope, and an appropriate cell is stained and photographed and the photograph is enlarged. Individual chromosomes are then cut out and arranged in pairs.

A *karyotype* (Fig. 26.2) is the arrangement of the chromosomes of a single cell in pairs. The autosomes are arranged in pairs according to:
(a) the length of the chromosome,
(b) the position of the centromere.

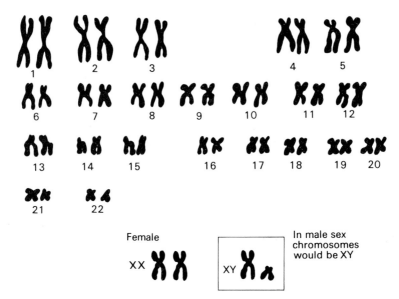

Fig. 26.2. Chromosomes displayed in pairs in a 'karyotype'. The sex chromosomes are below.

They are then arranged in seven groups A–G. A is of the longest chromosomes, G of the smallest. The pairs are numbered 1–22. Several of the pairs within a group look identical and cannot be separately distinguished with certainty by visual means.

Sex chromatin

A *Barr body* (Fig. 26.3) is a small mass of sex chromatin formed of a fusion of two X chromosomes. It can therefore be only found in women. When present it lies against the membrane of the nucleus of a cell. It is demonstrated most readily in the cells of a smear obtained from scraping slightly the mucous membrane of the mouth. About 20 per cent of the epithelial cells in a buccal smear from a woman show a Barr body.

In about three per cent of polymorphs in women the sex chromatin projects from the nucleus like a drumstick.

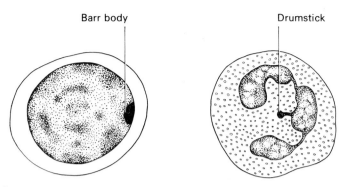

Fig. 26.3. A Barr body found in female cells. A 'drumstick' found in some female granulocytes.

A Barr body and a drumstick are not found in males because it is necessary to have two X chromosomes to form them, and males have only one X chromosome. The presence or absence of Barr bodies can be used to determine whether a person is genetically a male or a female when there is doubt, e.g. in some athletes.

Meiosis

Meiosis (not mitosis) is the special form of division of sex cells, for in the reproduction of sex cells, each unit (ovum or sperm) must finish up with half the usual number of chromosomes.

An ovum must have 22 autosomes and one X cell.

A sperm must have 22 autosomes and one X cell or one Y cell. This halving is achieved at one of the stages of division.

In the fertilization of an ovum:

either

(a) an ovum with 22 autosomes and 1 X cell is fertilized by sperm with 22 autosomes +1 X cell; and a female child is produced,

or

(b) an ovum with 22 autosomes and 1 X cell is fertilized by a sperm with 22 autosomes +1 Y cell, and a male child is produced.

The sex of the child depends upon whether the sperm which does the fertilizing carries an X or a Y chromosome.

Mutation

A mutation is a change produced in the structure of the DNA of a parent cell so that daughter cells inherit a pattern slightly different from that of the parent cells. Mutation may occur as a result of:

(a) exposure to radiation, e.g. the ordinary background radiation to which everyone is exposed,

(b) changes in temperature,

(c) certain drugs used to stop cancer cells from multiplying.

Some mutations are beneficial to the organism and tend to become established. Other mutations are harmful; they may be incompatible with life and kill the embryo or they may produce disease. The sudden appearance of an inherited disease in a family previously unaffected by it may be the result of such a mutation. With the continual appearance of genetic changes, good and bad, the population of a country acquires 'genetic equilibrium'.

Clinical features
Chromosomal abnormalities are common. Probably one fetus in ten is affected. Many of the abnormalities being incompatible with life, most of the affected fetuses die in utero and are spontaneously aborted.

About one in 200 live-born children has a chromosomal abnormality. The abnormality may be of two types.
1. An *excessive number of autosomes*. In *Down's syndrome* (mongolism) each cell has 47 chromosomes, there usually being three chromosomes no. 22 instead of the normal two; the result of the abnormality is mental retardation and an abnormal physical development.

Translocation is the attachment to one chromosome of a piece broken off another. *Mosaicism* is the condition in which some but not all the cells of a body show a chromosome abnormality.

2. An *abnormality of the sex chromosomes*:
(a) *Klinefelter's syndrome*: 2 X and 1 Y chromosome are present. Affected people are males (because of the Y chromosome), tall, thin and sexually underdeveloped.
(b) *Turner's syndrome*: only 1 X chromosome. Affected people are women (no Y chromosome), short, with a webbed neck and a failure of ovarian function.
(c) *Super-female*: 3 X chromosomes; affected women may have amenorrhoea and are mentally retarded.

Inheritance of diseases and characteristics

> *Types of genes*
> dominant
> recessive
> sex-linked

DOMINANT GENES

If the chromosomes of either parent contains a dominant gene, this gene is transmitted to half of his or her children. This half will show the particular feature produced by the gene and will transmit it to half their children.

The other half will not inherit the dominant gene and cannot suffer the effects nor transmit it.

The degree to which a dominant gene can affect a person varies. Genes can be affected by other genes and by the environment. The 'expression' of a gene (i.e. its ascertainable effects) may be full or decreased. If the 'expression' is much reduced in a person, the gene appears to 'skip' a generation, possibly to appear with full 'expression' in the next generation.

Clinical features
Conditions produced by dominant genes include:
 achondroplasia: a form of dwarfism produced by a fault in bone formation from cartilage.
 Huntington's chorea: a combination of dementia and involuntary movements.
 dystrophia mytonica: a disease of muscle and other organs.

RECESSIVE GENES

To produce effects a recessive gene must be present in both parents. Any child of such parents has a one in four chance of inheriting the condition.
 If only one parent carries a recessive gene, any child of his or hers will inherit the gene without showing any effects and will transmit it to his or her children. The chance of inheriting the same recessive gene from both parents is much increased if the parents are cousins.

Clinical features
Conditions produced by recessive genes include:
 phenylketonuria: a metabolic disorder produced by a biochemical defect with an accumulation of phenylalanine in the blood and mental retardation.
 Wilson's disease (hepato-lenticular degeneration): a disease of copper metabolism, affecting in particular the brain and the liver.
 sickle-cell anaemia: a form of anaemia in which the shape of red cells is distorted.

SEX-LINKED GENES

Conditions produced by sex-linked genes are usually transmitted on an X chromosome. The condition is transmitted by females who are not usually affected and expressed in males.

Clinical features
Conditions produced by a sex-linked recessive gene include:
 colour blindness,
 haemophilia: a disease characterized by bleeding,
 pseudo-hypertrophic muscular dystrophy: a degeneration of muscle.

27

Development before Birth

A new human being is created by the fertilization of an ovum by a spermato-zoon (Fig. 27.1)

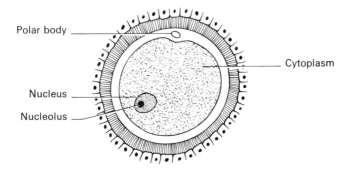

Fig. 27.1. An ovum and spermatozoon.

Ovum

At birth the ovaries contain several thousand immature ova. From puberty to the menopause about one ovum a month becomes mature. Those which do not mature gradually atrophy. A mature ovum is a cell about $\frac{1}{6}$ mm in diameter, just large enough to be visible to the naked eye against a dark background.

The ovum is contained within a Graafian follicle (Fig. 27.2). A *Graafian follicle* consists of a fluid-containing cavity lined by a layer of cells called the membrana granulosa, outside which are two further layers. The ovum develops within the membrana granulosa. With the maturing of the ovum the Graafian follicle becomes visible as a projection on the surface of the ovary, and even-tually its outer surface breaks down and the ovum is discharged.

An ovum consists of:

(a) a zona pellucida, a membrane enclosing the rest of the ovum,
(b) cytoplasm, containing a small amount of yolk,
(c) a nucleus, inside which is a nucleolus.

The nature ovum is discharged from one or other ovary in each menstrual cycle, about 14 days after the beginning of the preceding menstrual flow. The

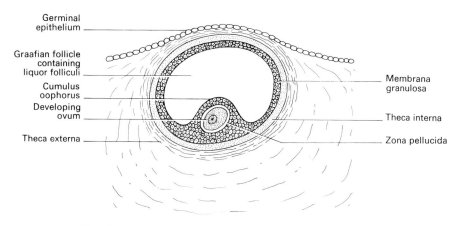

Germinal
epithelium

Graafian follicle
containing
liquor folliculi

Cumulus
oophorus

Developing
ovum

Theca externa

Membrana
granulosa

Theca interna

Zona pellucida

Fig. 27.2. An ovarian follicle on the surface of the ovary.

fimbriae at the end of the uterine tube grasp the ovary, and the ovum passes through the opening in it. Once inside the tube, the ovum is propelled along by muscular contractions of the wall of the tube and by the sweeping movements of the ciliated cells of its mucous membrane. It starts to degenerate if it is not fertilized by a sperm within about 24 hours.

Spermatozoon

Spermatozoa (sperm) start to be produced at puberty and continue to be produced at the rate of several millions a day for the rest of the man's life. A sperm is much smaller than an ovum. It consists of a head (containing the nucleus of the cell), a neck, and a long tail, which by wagging moves the sperm along at a rate of about 1 mm per minute (Fig. 27.3). The fertilizing power of sperm is lost in about 24 hours, but it can be preserved by freezing them in an appropriate medium.

Fertilization

In sexual intercourse about 200 million sperm in about 3.5 ml of semen are deposited at the top of the vagina or within the canal of the cervix of the uterus. Some of the sperm move up the uterus, partly by their own movements, partly aided by muscular contractions of the uterus, and a few hundred get as far as the uterine tube. A sperm cannot fertilize an ovum until it has been exposed to uterine secretions for about 8 hours. Up to a hundred sperm are likely to reach an ovum. One of them enters the ovum by penetrating the outer layer, and as soon as this happens a chemical change occurs in the layer which prevents the entry of other sperm.

The nucleus of the ovum and the nucleus of the sperm fuse. The rest of the

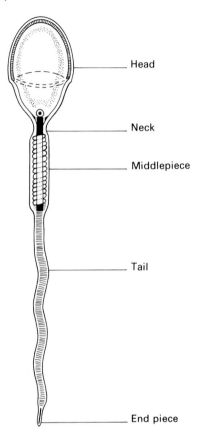

Head

Neck

Middlepiece

Tail

Fig. 27.3. A spermatozoon.   End piece

sperm atrophies. The fertilized ovum immediately begins to divide—into two cells, into four and so on. The yolk in the cytoplasm is used up in the process and the zona pellucida becomes thin and disappears.

The ovum continues to move along the uterine tube and reaches the uterine cavity about the 3rd–4th day after fertilization. For another three or four days it remains free at the top of the uterine cavity, drawing its oxygen and nourishment from the uterine secretions surrounding it. It then embeds itself in the endometrium. At this stage of its development it is called a blastocyst.

Under the influence of progesterone, a hormone secreted by the corpus luteum of the ovary, the endometrium has become thick and vascular and ready to receive the blastocyst.

Blastocyst

The *blastocyst* is round with an outer wall of cells called the *trophoblast* enclosing a fluid-containing cavity with a clump of cells called the *inner cell mass* projecting into the cavity at one point. The inner cell mass sinks into the endometrium, an action which is called the implantation and usually occurs high up on the posterior wall of the uterus. The endometrium then grows over the inner cell mass.

The *embryo* develops from the inner cell mass.

Contact with the mother is made through the cells of the trophoblast, through which pass the substances necessary for the development of the embryo. Part of the trophoblast develops into the placenta.

Fig. 27.4. A fetus at three weeks.

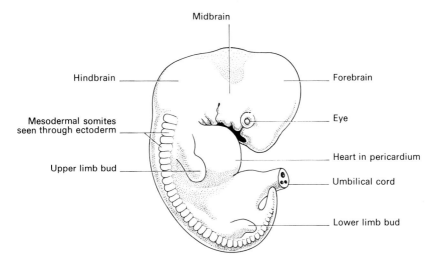

Fig. 27.5. A fetus at four weeks.

Development of the embryo

Rapid growth and a number of complicated changes take place which result in the formation of:

the *embryo* (Figs 27.4, 27.5).

the *membranes*, the amnion and the chorion, which fuse together to enclose a fluid-containing sac. This sac contains the embryo. The outer wall of the sac enlarges until it occupies the whole of the uterine cavity.

the *placenta*, through which gaseous and chemical exchanges take place.

a *stalk* connecting the embryo to the placenta.

After three months the *embryo* is called a *fetus* (the spelling 'foetus' is incorrect).

Ectoderm, mesoderm, endoderm

The embryo develops three basic tissues—the ectoderm, the mesoderm and the endoderm—from which all the tissues and organs of the body are derived.

Ectoderm
  The epidermis of the skin and its derivatives—
      hair, nails, sweat glands, sebaceous glands,
      breasts
  nervous system and its derivatives—posterior
      lobe of the pituitary gland, pineal gland,
      medulla of the adrenal gland, retina
  epithelium of mouth, nasal cavities, anal canal
  lens of eye

---

mesoderm
  bone, cartilage, joints
  connective tissue
  muscle—skeletal,
      smooth, cardiac
  heart   blood   blood   spleen
      vessels
  lymph nodes and   cortex of adrenal gland
      vessels
  urinary system   reproductive system
      (mostly)   (mostly)

---

endoderm
  alimentary system   Liver   pancreas
  respiratory system
  thyroid gland   parathyroid glands
  thymus (partly)

DEVELOPMENT OF ORGANS AND TISSUE

Skin

The epidermis of the skin and its derivatives develop from ectoderm. The dermis develops from mesoderm. The hair develops as buds of cells which grow downwards and from the bottom of each bud grows the shaft of a hair. Sebaceous glands develop as lateral outgrowths from the hair-buds. Sweat glands develop as downgrowths in the skin about the 4th month of fetal life. Nails appear about the 3rd month. The breasts develop as modified sweat glands.

The *vernix caseosa* is a mixture of sebaceous secretion and desquamated skin which forms a white slippery coat over the skin from the 6th month onwards.

Nervous system

The nervous system develops from ectoderm. During the 3rd week a *neural plate* develops as a thickening of the ectoderm along the back of the embryo, from head to tail. The plate grows downwards, into the embryo, and becomes grooved. The edges of the groove become fused together to form a tube called the *neural tube* which becomes separated from the skin of the back of the embryo.

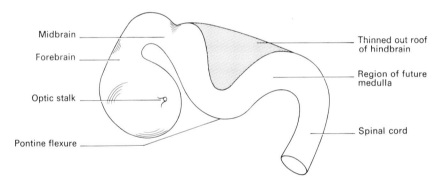

Midbrain

Forebrain

Optic stalk

Pontine flexure

Thinned out roof of hindbrain

Region of future medulla

Spinal cord

Fig. 27.6. The early development of the brain.

The *brain* (Fig. 27.6) develops as the specialized head-end of the neural tube. The forebrain, midbrain and hindbrain develop as three enlargements, and some bends appear at the head-end. The *forebrain* grows into the two cerebral hemispheres, by far the largest part of the brain and overlapping the rest. The thalami and basal ganglia appear in it, and the motor nerves grow out of cells in the developing cortex and start to grow downwards towards the spinal cord. By the end of the 7th month all the main sulci have appeared. The *midbrain* grows least and becomes only a relatively small amount of the fully developed brain. The *hindbrain* grows into the pons, medulla oblongata and cerebellum.

Neurones develop at a rapid rate (many thousands a minute at one stage), migrate to other parts of the nervous system, and succeed in connecting with the axons and dendrites of other neurones in precise patterns. Unlike most cells, neurones do not divide once they are formed; once dead they are not replaced, although sometimes (especially early in life) their functions are taken over by other neurones.

The *cranial nerves* develop from the brain. The *nervous parts of the eye* and the *posterior lobe of the pituitary gland* (Fig. 27.7) are outgrowths from the forebrain.

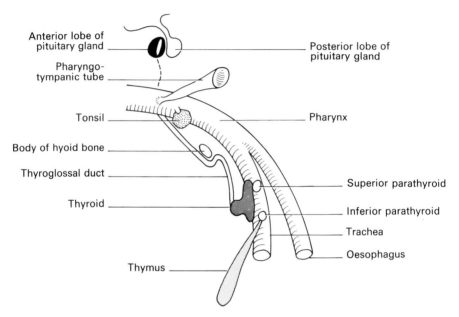

Fig. 27.7. The origins of some structures in the neck.

The *spinal cord* develops from the rest of the neural tube. The *spinal nerves*, the *sympathetic nervous system* and the *medulla of the adrenal glands* are all outgrowths from it.

In the early embryo the spinal cord extended the whole length of the spinal canal within the developing spine; but as it does not grow as fast as the rest of the embryo, its tail-end gradually becomes withdrawn up to the upper lumbar region. The lower spinal nerves thus come to pass more and more obliquely down the spinal cord, the lowest ones forming the cauda equina.

The *ventricles* of the brain and the tiny canal in the grey matter of the spinal cord are developments from the original cavity that ran the length of the neural tube.

Clinical features
The sides of the neural tube may fail to fuse. *Anencephaly* is the result of a failure to fuse at the head-end: the brain does not form and the child dies shortly after birth. *Spina bifida* is the result of a failure to fuse at the tail-end, the failure

of fusion of nervous tissue being associated with a failure of fusion of the mesodermal tissues that are forming round the tube to make the vertebrae.

Alimentary tract

The alimentary tract develops from endoderm, except for the epithelium of the mouth and the lower end of the anal canal, which are ectodermal ingrowths from the skin.

The *gut* is a tube of endoderm pinched off from the yolk sac by the development of the head- and tail-ends of the embryo. Both ends of the tube are at first closed.

> *Gut*
> foregut
> midgut
> hindgut

FOREGUT

> *Structures formed from foregut*
>
> | | |
> |---|---|
> | mouth | larynx |
> | nose | trachea |
> | pharynx | bronchi |
> | oesophagus | lungs |
> | stomach | thyroid gland |
> | duodenum (in part) | parathyroid glands (in |
> | liver | part) |
> | pancreas | thymus (in part) |

The blind front end of the foregut lies between the rapidly developing brain and the rapidly developing heart. At this point the ectoderm covers a little pit, at the bottom of which is the membrane that closes off the foregut. About the 4th week this membrane disappears with the formation of a mouth opening into the foregut.

Three *palatal processes* then grow across this opening, two large lateral ones and a small central one. These processes are bars of mesoderm, covered on the outside with ectoderm and on the inside with endoderm. By their fusion they form the *upper lip* and the *palate*, and they separate thus the nose from the mouth.

Lower down, below the mouth, mandibular processes grow from either side to fuse in the middle and form the *lower jaw* (Fig. 27.8).

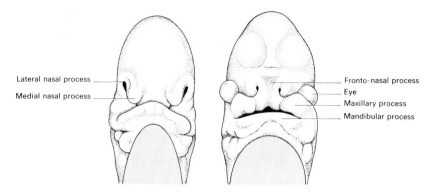

Fig. 27.8. The development of the jaws.

Clinical features

Hare-lip and *cleft palate* are caused by a failure of development and fusion of the palatal processes. A *hare-lip* is usually to one side or other of the midline for the commoner failure is that of one lateral process to fuse with the central process. A central hare-lip is a grosser deformity resulting from the failure of both lateral processes to fuse with the central process. A *cleft palate* can be of any degree between a complete cleft of hard and soft palates (often associated with a hare-lip) and a division of the uvula into two.

A *cleft chin* is an indication of a failure of fusion of the mandibular growths. In its minimal form it appears as a dimple in the midline.

The five *branchial arches* are bars of mesoderm, lined on the outside with ectoderm and on the inside with endoderm, which develop in the lateral region of the pharynx. From them develop skeletal elements—including part of the mandible, the hyoid bone, the cartilages of the larynx—and associated muscles. The *tongue* is formed from the 1st and 2nd of these arches, the sulcus terminalis on its surface marking the division between these two origins. As the branchial arches develop, four *pharyngeal pouches* of endoderm project laterally between the arches. The thymus and the thyroid gland in part and all the parathyroid glands develop from these pouches.

The *oesophagus* develops as a simple tube, elongated by the growth of the thorax. The *stomach* appears at the 4th–5th weeks as a dilatation of the gut. At first it lies anteroposteriorly, but as it develops it rotates through a right angle so that its right side now faces backwards and its left forwards. Its left side grows faster than its right and forms the big bulge of the greater curvature.

The *liver* and the *pancreas* with their ducts develop as outgrowths from the foregut just beyond the stomach, and the courses their ducts take show the lines along which they have grown.

Clinical features

A malformation of the developing lung-bud (which grows out of the foregut) can cause the oesophagus to be divided into two parts. The upper part opens into the pharynx above and ends blindly about the level of the tracheal bifurcation. The lower part opens into the trachea above and into the stomach below.

MIDGUT

> **Structures formed from midgut**
>   duodenum (in part)
>   ileum
>   jejunum
>   caecum
>   appendix
>   large intestine (as far as sigmoid colon)

The gut here grows in length more than the embryo and forms a big loop forwards. Smaller loops of small intestine develop as the midgut continues to grow rapidly in length, although being at present functionless it remains thin.

The *caecum* is formed as a bulge at the beginning of the large intestine. The *appendix* begins as the pointed tip of the caecum. As the right side of the caecum develops more quickly than the left and comes to form the greater part of the caecum, the appendix is left as a remnant on its left side.

Rotation of the gut occurs. In this action the caecum and ascending colon move over to the right side of the abdomen across the front of the small intestine, dragging their peritoneal attachments with them. The movement brings the caecum, ascending colon and transverse colon into their permanent positions.

Meckel's diverticulum, a blind-ended projection from the ileum, represents a partially persisting stalk. A fibrous cord extending from the ileum to the umbilicus is a persistent remnant of the stalk.

HINDGUT

> **Structures formed from hindgut**
>   rectum
>   upper end of anal canal
>   bladder
>   urethra

The *cloaca* is the dilated bottom end of the hindgut. It is closed off below by the *cloacal membrane*, formed of endoderm on the inside and ectoderm on the outside.

The cloaca is divided into a front part and a back part by a septum which grows down from above and fuses below with the cloacal membrane.

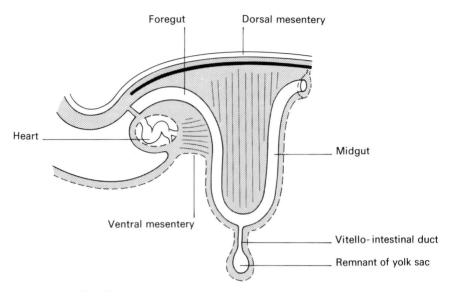

Fig. 27.9. An early stage in the development of the gut.

The *front part* of the cloaca becomes bladder and urethra.

The *back part* of the cloaca becomes the rectum and the upper end of the anal canal.

The two parts of the cloacal membrane formed by union of septum and membrane now break down to form openings to the exterior: the opening of the urethra in front, the opening of the anus behind.

Clinical features

A *fistula* forms between the bladder and rectum if the septum dividing the cloaca into two is not perfect. An *imperforate anus* is the result of a failure of the posterior part of the cloacal membrane to break down.

Cardiovascular system

The cardiovascular system is formed from mesoderm. A number of tubes develop and link to form a primitive circulatory system. Then the heart and various large blood vessels are formed, and a functioning circulatory system appears, the essential parts of which are:

(a) the developing heart, which acts as a pump for the system,

(b) a primitive aorta beginning at the head and running backwards,

(c) two umbilical arteries connecting the aorta (later via the internal iliac arteries) with the developing placenta through the umbilical cord,

(d) the placenta attached to the endometrium of the uterus,

(e) two umbilical veins (later one) running from the placenta via the umbilical cord,

(f) a primitive major vein which receives the blood from the umbilical veins and transports it to the heart.

THE HEART

The *heart* (Fig. 27.10) is formed from a large vessel immediately under the foregut, and makes a big bulge in the front of the embryo.

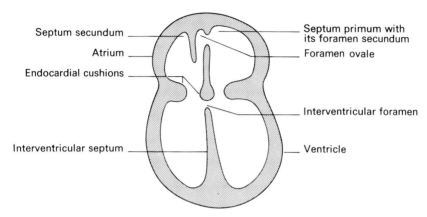

Septum secundum

Atrium

Endocardial cushions

Interventricular septum

Septum primum with its foramen secundum

Foramen ovale

Interventricular foramen

Ventricle

Fig. 27.10. An early stage in the development of the heart, with the foramen ovale between the atria and the interventricular foramen.

At an early stage it consists of:
(a) the sino-atrial chamber, into which the veins open,
(b) the right ventricle,
(c) the left ventricle,
(d) the truncus arteriosus opening out of the left ventricle.

At first these chambers are arranged in a line, one after the other. In the course of a series of complicated changes, the chambers of the heart, their valves and the origins of the great vessels develop. As the position of the heart is fixed, it cannot in making these changes remain straight but develops various bends. At first, the right and left atria develop at the venous end of the heart, out of the sino-atrial chamber. To separate the two atria, two septa grow towards one another but do not fuse, leaving a flap-like valve called the *foramen ovale*, through which blood can pass from the right atrium to the left but not in the opposite direction. The tricuspid and mitral valves form out of endocardium.

Secondly, an interventricular septum grows to separate the two ventricles, and thence spirally upwards into the truncus arteriosus.

Following this the truncus arteriosus becomes divided by this septum into aorta and pulmonary artery, with the aorta opening out of the left ventricle and the pulmonary artery out of the right. At the junction of ventricle and great vessel, endocardial cushions grow inwards and become hollowed out into three cusps for the aortic valve and three for the pulmonary valve.

Finally, primitive cardiac muscle fibres develop in the walls of the chambers, especially those of the ventricles. The cardiac muscle starts to beat spontaneously, at first irregularly, later regularly, before any nerve has reached the heart. Later the coordination of the beat is taken over by the developing sino-atrial node and the conducting system.

AORTIC ARCHES

Bilateral intercommunicating *aortic arches* or arteries develop at the head-end. At first a right and left aorta are present, but they fuse lower down to become one vessel. In the course of development several of these arches disappear. Most of the right aorta disappears—a length of it persisting as the right subclavian artery. The left aorta persists as the arch of the aorta and upper part of the descending thoracic aorta.

The *ductus arteriosus* is a part of an arch which connects the left pulmonary artery with the aorta. It is an essential artery in the fetal circulation.

Clinical features

Various congenital abnormalities can occur in the heart. In *dextro-rotation* the heart and great vessels are a mirror-image of the normal; intra-abdominal organs are also reversed.

Fallot's tetralogy is the result of a failure of the division of the truncus into aorta and pulmonary artery. The ventricles communicate with one another, and the pressure inside the right is as high as that in the left. The tetralogy consists of:

(a) a ventricular septal defect,
(b) a narrow pulmonary artery,
(c) an 'over-riding' aorta, which is too big and receives blood from both ventricles,
(d) right ventricular hypertrophy.

As venous blood enters the aorta, one result is a 'blue baby'.

Respiratory system

The larynx, trachea, bronchi and lungs (Fig. 27.11) develop from a diverticulum which appears in the floor of the pharynx and grows downwards, separating itself from the oesophagus immediately behind it. The lower end of the diverticulum splits into two lung-buds, in which develop the bronchial trees and alveoli of each lung. Blood vessels move into the growing lungs to form the pulmonary vessels.

At the beginning of the 6th month the original endoderm is shed off from the alveoli, and from this stage onwards the lungs are capable of functioning and the fetus is viable. But before birth the alveoli are small and contain swallowed amniotic fluid.

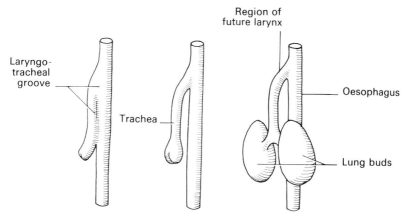

Fig. 27.11. The growth of the trachea and lungs from the pharynx.

The diaphragm develops for the most part in mesoderm in the neck, just under the developing heart. As the heart and lungs develop, the diaphragm is pushed down until it reaches its permanent position. As it is pushed down, it drags with it the nerve supply it received while it was in the neck. This is why the nerves to the diaphragm are the phrenic nerves, which have their origins in cervical nerves 3–5.

Urinary system

The urinary system is formed mainly from mesoderm.

The *urethra* and *bladder* are formed from the cloaca of the hindgut and are therefore formed of endoderm. The *ureter* grows outwards from the bladder-to-be and ends in a little bud from which is formed the *pelvis of the kidney* at the top end of the ureter and the *collecting tubules*. The rest of the *kidney* forms in mesoderm in the pelvis, just below the bifurcation of the aorta. The distance to be traversed by the developing ureter is therefore not far. The kidneys then migrate upwards along the posterior abdominal wall to their permanent position just below the adrenal glands. There they acquire a new blood supply from the aorta. At birth the kidneys still show an early lobulated form.

Clinical features
A kidney may stop anywhere along its line of ascent. A *horseshoe kidney* is due to fusion of the lower poles of the kidney as they start the ascent. The *ureter* may be double for part of its length. An *abnormally placed renal artery* may obstruct the ureter and cause hydronephrosis. *Polycystic disease of the kidney* is the result of a failure of some nephrons to connect with the collecting tubules growing towards them. Cysts are formed as the tubules become distended with secretion. A large number of cysts will impair renal function. *Ectopia vesicae* is a deformity in which the bladder opens on to the abdominal wall as the result of a split in the cloacal membrane which should have closed off the front of the bladder.

MALE GENITAL SYSTEM

The *testis* develops from mesoderm high up on the posterior abdominal wall and becomes connected with the developing epididymis and vas deferens. The *scrotal swellings* develop between the thighs. The *processus vaginalis* is a projection of the peritoneal cavity down into the scrotal swelling on each side.

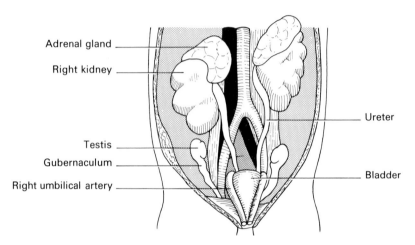

Adrenal gland

Right kidney

Testis

Gubernaculum

Right umbilical artery

Ureter

Bladder

Fig. 27.12. The early development of the male genito-urinary tract.

As the testis enlarges, it descends along the gubernaculum. The *gubernaculum* is a cord of condensed tissue extending from the testis to the scrotal swelling and by contracting draws the testis downwards in the following time-table of fetal life (Fig. 27.12):

3rd month: at the iliac fossa,
7th month: in the inguinal canal,
8th month: at the lower end of the inguinal canal,
9th month: in the scrotum.

Clinical features
An *undescended testis* is one that has failed to descend the whole course and remains somewhere along the route—within the abdomen, within the inguinal canal or at the upper end of the scrotum. A testis retained within the abdomen cannot withstand the core heat and fails to produce sperm.

FEMALE GENITAL SYSTEM

The *ovary* develops from the same basic tissue as the testis. Primitive germ cells form in it to become the ova.

The *uterine tubes*, *uterus* and *vagina* develop from two ducts called the Mullerian ducts. The upper end of each duct remains separate and forms a uterine tube. The lower ends fuse together to make at first a solid bar of tissue.

Later this bar becomes canalized throughout its length, and its lower end opens between two genital swellings which become the labia majora and minora.

Clinical features
Developmental abnormalities that can occur include:
(a) absence or failure of proper development of the Mullerian duct on either or both sides,
(b) persistence of a double Mullerian duct for all or part of the way, with the production of a double uterus or vagina,
(c) failure of the duct to break down between the genital swellings, so that a complete hymen blocks the lower end of the vagina and prevents the escape of secretions and menstrual fluid.

PITUITARY GLAND

The pituitary gland is formed in two parts. The *anterior lobe* is formed by an upgrowth from the primitive nose-mouth area, its connection with that area being lost. The *posterior lobe* is formed by a downgrowth from the floor of the forebrain; the stalk connecting it to the forebrain persists as the stalk of the pituitary gland.

THYROID GLAND

The thyroid gland is formed mainly as a downgrowth from the floor of the pharynx; the lateral lobes of the gland are formed partly from the 4th pharyngeal pouches.

Clinical features
Thyroid tissue may develop at any site on the line from the foramen caecum to the gland. A *lingual thyroid* is one present at the base of the tongue. Sometimes the gland descends further than it should to become a *retrosternal thyroid gland*, i.e. one situated behind the sternum. A *thyroglossal duct* may be present along the line of descent of the gland. If the duct is closed, a cyst can form. The duct may open as a tiny hole in the midline of the neck.

ADRENAL GLANDS

The adrenal glands are formed from two sources. The *cortex* is formed from mesodermal tissue lying at the back of the abdomen. The *medulla* is formed from nervous tissue and is essentially part of the sympathetic nervous system. The two parts lie side by side for a time, and then the cortex encloses the medulla. The gland is relatively large in the fetus.

Muscle, skeleton, limbs

In early embryonic life the body becomes segmented into a number of segments of muscle each with its own nerve and artery. This segmented pattern is disturbed by the development of the head and limbs, but the pattern can be seen in:

vertebrae   spinal nerves

ribs   intercostal vessels and nerves.

A *myotome* is one of the muscle segments from which develop the muscles of the trunk.

Limb-buds grow out of the trunk about the 4th week. Those for the arms are a little in advance of those for the legs. At first they consist of mesoderm covered with ectoderm. The nerve supply to the limbs is an indication of the segments from which they arise:

arm: cervical 4–8,

        thoracic 1–2,

leg: thoracic 12,

        lumbar 1–4.

Each limb-bud develops in four segments: an upper segment, a lower segment, a hand or foot segment, a digit segment. Appropriate cartilages for the bones develop in the segments, with joints forming between them. Periosteum grows around the cartilage, and bone starts to form from the inner surface of the periosteum and from a primary centre of ossification in each shaft.

Nerves from the appropriate segments grow into the developing limb-bud. An *axis artery* develops in each limb as its main artery. In the arm this becomes the subclavian–axillary–brachial artery, with the radial and ulnar arteries developing later. In the leg the axis artery is at first at the back of the thigh close to the sciatic nerve and is continued as the popliteal and posterior tibial arteries; but the functions of the axis artery at the back of the thigh is later taken over by the femoral artery, which develops at the front of the thigh.

Placenta

The *placenta* (Fig. 27.13) is the organ through which the maternal and fetal interchange take place. It acts as the fetal lungs, alimentary tract and kidneys, and is fully active from the 4th month onwards.

When fully formed it is disc-shaped. On the amniotic side its surface is smooth. On the uterine side its surface is rough, being marked by several depressions into segments called cotyledons; this surface is embedded in the

---

Functions of placenta
    exchange between maternal and fetal blood of:
        oxygen
        carbon dioxide
        foodstuffs
    production of hormones (oestrogen,
        progesterone, gonadotrophins) essential for
        development of uterus and breast during
        pregnancy

Substances passing the placental barrier
The placenta acts as a barrier which allows
 certain substances to be exchanged between
 maternal and fetal blood
The following can pass through the barrier:
oxygen | carbon dioxide
water | amino acids
fatty acids | glucose
vitamins | sodium, potassium and other
 salts
some drugs | some antibodies and antigens
some micro-organisms (e.g. German measles
 virus)

endometrium of the uterus. The umbilical cord is usually attached to it about its middle. Occasionally, in a velamentous insertion, the blood vessels spread out in the membranes before reaching the placenta.

The blood vessels of the umbilical cord end in a capillary network in the numerous villi which project from the cotyledons into the endometrium. The maternal blood circulates in the spaces around the placental villi.

Fetal and maternal blood do not come into contact.

Umbilical cord

The umbilical cord, when fully developed, is about 50 cm long. It connects the

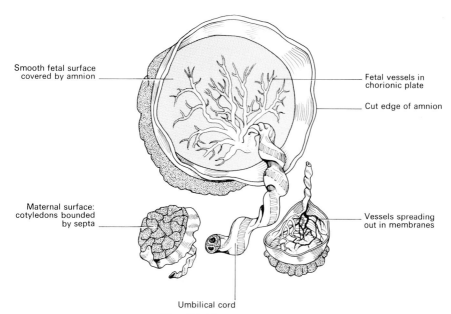

Fig. 27.13. The placenta.

fetus to the placenta. The vessels in it are twisted, always in the same direction. It is composed of:

an umbilical vein,

two umbilical arteries,

Wharton's jelly, the connective tissue surrounding the vessels.

Amniotic cavity

The amniotic cavity is the fluid-occupied cavity in which the embryo–fetus develops. The *amniotic fluid* which occupies it is composed of:

fluid secreted by the amnion,

urine from the fetus when functioning kidneys have developed and the cloacal membrane has broken down and allows it to escape.

At term (i.e. the 9th month) the amniotic fluid amounts to about 700 ml.

Membranes

The membranes are the amnion and the chorion, which become fused together.

The *amnion* lines the surface of the cavity nearer to the fetus and secretes amniotic fluid.

The *chorion* started off with a number of villi which projected into the endometrium, but with the growth in size of the amniotic cavity these villi disappear except at the area where the placenta develops.

Fetal circulation

The circulation before birth is different from the circulation after birth because before birth the lungs do not function and gaseous exchanges take place in the placenta (Fig. 27.14).

1. Oxygenated blood passes from the placenta into the umbilical vein.

2. The umbilical veins pass along the umbilical cord and from the umbilicus ascend to the inferoposterior surface of the liver.

3. Here the oxygenated blood passes into the inferior vena cava either

(a) directly through a vessel called the ductus venosus, or

(b) indirectly through the sinusoids of the liver and thence into the hepatic veins which enter the inferior vena cava at the back of the liver.

4. The blood in this upper section of the inferior vena cava is a mixture of

(a) oxygenated blood from the placenta,

(b) non-oxygenated blood from the lower half of the fetus.

5. This mixture of blood enters the right atrium of the heart.

6. The right atrium receives blood from two sources:

(a) the blood from the inferior vena cava,

(b) non-oxygenated blood from the fetal head and arms through the superior vena cava.

7. The partly oxygenated blood from the inferior vena cava passes through the

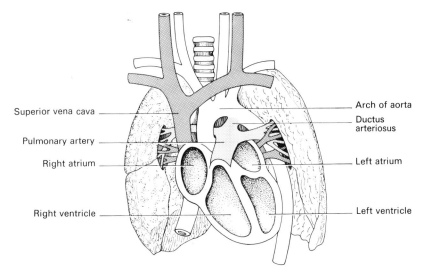

Fig. 27.14. The fetal heart with the ductus arteriosus connecting the pulmonary trunk with the aorta. The lungs are not expanded.

foramen ovale, the opening between the two atria, and so enters the left atrium. A little blood also comes into the left atrium from the lungs through the pulmonary veins. From the left atrium this blood passes into the left ventricle and is expelled into the aorta by ventricular contraction. Some of this blood goes to the head and arms and is returned to the heart through the superior vena cava.

8. The blood coming into the right atrium through the superior vena cava is directed by a septum through the tricuspid valve into the right ventricle. It is driven by ventricular contraction into the pulmonary artery.

About one-fifth of this blood goes through the lungs and thence into the left atrium. Four-fifths pass through the ductus arteriosus, which connects the left pulmonary artery with the arch of the aorta immediately above it. This blood, together with some that has come down the arch of the aorta, passes into the branches of the abdominal aorta to supply the abdominal viscera and the legs.

9. From each internal iliac artery an umbilical artery passes upwards and inwards, behind the anterior abdominal wall, to reach the umbilicus. The two umbilical arteries, carrying non-oxygenated blood, run in the umbilical cord to the placenta.

10. Gaseous interchanges of oxygen and carbon dioxide take place between the fetal blood in the villi of the placenta and the maternal blood in the spaces surrounding the villi.

The foramen ovale may remain open as a way between the two flaps that form it. In the absence of complicating factors, this defect is symptomless.

The ductus arteriosus may remain open (Fig. 27.15). When this happens

> *Changes in the circulation at birth*
> 1. The muscle in the wall of the umbilical arteries contracts and pulsation in them ceases. The umbilical arteries within the body ultimately become fibrous bands.
> 2. The lungs become expanded with the first gasps the baby takes; the blood vessels in the lungs open up; respiration begins.
> 3. The foramen ovale closes because the pressure in the left atrium becomes greater than that in the right and the two flaps of which the foramen is made are pressed together and fuse within a few days.
> 4. The muscle in the wall of the ductus arteriosus contracts (probably as a reaction to chemical changes in the blood) and in the course of two weeks closes to become a fibrous band.

blood passes in the reverse direction because the pressure in the aorta is greater than that in the pulmonary artery. The ductus becomes dilated and thin-walled. This is a serious defect because infection of the ductus can occur and as a result of prolonged overloading of the left ventricle left ventricular failure and death can occur.

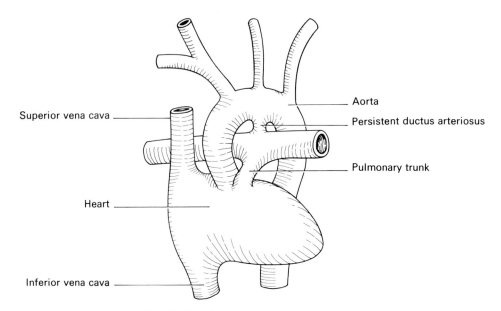

Superior vena cava

Aorta

Persistent ductus arteriosus

Pulmonary trunk

Heart

Inferior vena cava

Fig. 27.15. A persistent ductus arteriosus.

TWINS

Twins occur once in eighty births. The incidence of higher multiple pregnancies (triplets etc) is very low.

About three-quarters of all cases of twinning are *unlike (fraternal or dizygotic) twins* the result of the fertilization of two ova that happen to have been produced at the same time. The twins may be of either sex. They always have separate amniotic cavities and usually separate placentae.

Identical (monozygotic) twins have developed from the same fertilized ovum, most of them by a division into two of the inner cell mass at the blastocyst stage. According to the time at which the division takes place, they have separate or single placentae. Monozygotic twins are always of the same sex and are genetically and immunologically identical.

Clinical features

Conjoint twins are a rare congenital abnormality in which there is incomplete separation of twins which are anatomically joined; the join may be slight or there may be a mingling of organs between them in such a way that separation is impossible.

When twins share a placenta there is a risk that one will take more from the placenta than the other, which being deprived of nourishment becomes a weakling who may die before or soon after birth. The same effect may be produced when there are two placentae if one is more favourably placed in the endometrium than the other.

28

Development after Birth

The stages of life after birth are:
   infancy (the 1st year of life) and childhood,
   puberty and adolescence,
   adult life,
   old age.

INFANCY AND CHILDHOOD

Weight

There is a 5–6 per cent loss of weight due to fluid loss during the first few days of life. The birth-weight should be regained within 10 days.

The baby should double his birth-weight in six months and treble it in twelve months.

At birth the weight is usually 3–4 kg (6–9 lb).

> *Factors causing low birth-weight*
>   maternal genes (small women tend to have
>   small babies)
>   premature birth
>   multiple births (twins etc)
>   infections during prenatal life
>   low maternal intelligence
>   smoking by mother

The international definition of a premature or immature infant is: a live-born baby with a birth-weight of 2.5 kg ($5\frac{1}{2}$ lb) or less.

LARGE BABIES

Maternal or paternal diabetes mellitus may be the responsible factor.

Length

The body length at birth is about 50 cm. It should be doubled in 12 months.

Head and face

The head is relatively well developed at birth because of the degree of development of the brain and of the organs of special sense (Fig. 28.1).

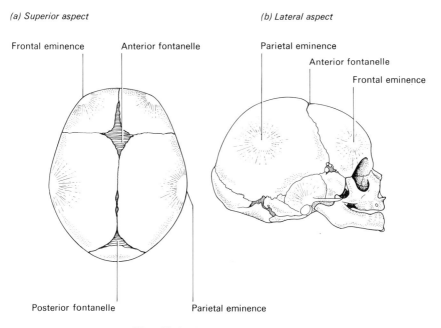

(a) Superior aspect

Frontal eminence   Anterior fontanelle

Posterior fontanelle   Parietal eminence

(b) Lateral aspect

Parietal eminence

Anterior fontanelle

Frontal eminence

Fig. 28.1. The skull at birth.

Circumference of head at birth: about 35 cm
                1 year: 45–46 cm
              2 years: 48–49 cm
              3 years: 49–50 cm

The rate of growth is related to the total growth of the child. A normal rapidly growing child has a rapidly growing head.

Causes of small head	*Causes of large head*
small baby	large baby
familial small heads	familial large heads
malnutrition	rapid growth
mental retardation	hydrocephalus
premature fusion of	
cranial bones	

The two *frontal bones* begin to unite at two years, and the union is complete at eight years. The *anterior fontanelle* (the membrane filling the gap between the two frontal bones and the two parietal bones) becomes ossified by three months.

The *posterior fontanelle* (the membrane filling the gap between the two parietal bones and the occipital bone) becomes ossified by eighteen months.

The *cranial sinuses* (maxillary, frontal, sphenoid, ethmoid) are barely present at birth. They enlarge slowly and do not reach full size until after puberty.

The *face* is relatively small at birth because:

(a) the teeth-bearing areas of maxilla and mandible (Fig. 28.2) are small,

(b) the maxillae are small because the antra have not yet developed,

(c) the nasal cavity is small.

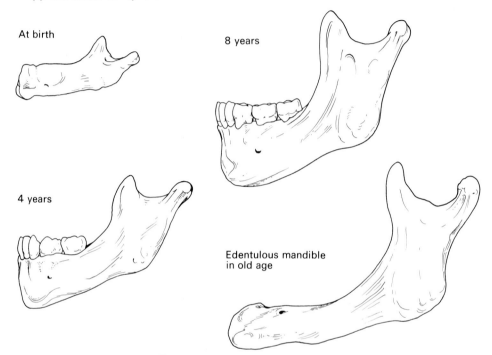

At birth

8 years

4 years

Edentulous mandible
in old age

Fig. 28.2. The mandible at different stages.

The *teeth of the first dentition* start to appear at six months and are all erupted by the twenty-fourth month. The *teeth of the second dentition* erupt between six and twenty-five years. The tip of the *tongue* is not developed at birth and the baby appears to be 'tongue-tied'.

Brain

Brain-weight at birth: about 300 g
           1 year: about 900 g
           6 years: about 1200 g

At six years the brain is fully grown.

The increase in the size of the brain during childhood is not due to any increase in the number of brain cells (they are all there at birth), but to myelination of the nerve fibres, an increase in the number of neuroglia, and an increase in the size of the blood vessels. Myelination of the corticospinal (pyramidal) tract is not completed until the 2nd year, and a child cannot walk until it has been completed.

With the increase in size of the brain, the smaller convolutions develop.

Heart

Pulse rate in newly born: about 130 per minute
at 1 year: about 110 per minute
at 5 years: about 100 per minute

The foramen ovale should close within a few days of birth and the ductus arteriosus within a fortnight.

Respiration

Respiratory rate in newly born: about 40 a minute
at 1 year: about 30 a minute
at 5 years: about 25 per minute

New alveoli continue to form at the ends of the air passages up to the 8th year.

Endocrine glands

Some of the endocrine glands are producing their hormones at birth. The *thymus* is large, extending at birth from the lower border of the thyroid cartilage to the upper border of the pericardium; it increases in size up to about the 8th year and thereafter diminishes. The *cortex of the adrenal gland* is large at birth.

Abdomen

The *kidneys* are still lobulated at birth. The bladder lies within the abdomen at first because the pelvis is too small to contain it, and its anterior wall is in contact with the anterior abdominal wall; it sinks gradually into the pelvis over several years. The *prostate gland* is rudimentary at birth.

The *testes* should be in the scrotum at birth.

The *cervix of the uterus* is longer than the body during infancy and child-hood.

Skin

The newborn baby is well covered with subcutaneous fat. The infant's chubby cheek is due to the presence in it of a pad of fat called the buccinator or sucking pad: it performs no function in sucking.

The lanugo, the fine hair which covers the body at birth, is replaced by vellus hair, which is inconspicuous, within a few months, and the fine hair on the scalp by thicker hair.

Skeletal system

At birth the shafts of the long bones are present as bone (Fig. 28.3), with cartilaginous ends. The only epiphysis that has appeared is that at the lower end

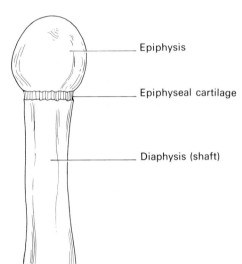

Epiphysis

Epiphyseal cartilage

Diaphysis (shaft)

Fig. 28.3. The end of a long bone before ossification of the epiphyseal cartilage.

of the femur. Other epiphyses continue to appear during childhood, and bony centres for the smaller bones.

Red bone marrow is present throughout the length of the shafts of the long bones up to the 4th year; but after that time more and more of the red marrow is replaced by fatty yellow marrow, a process which continues into puberty.

The *vertebral column* at first shows a concavity forwards (with a smaller concavity below of the sacrum). This upper concavity becomes changed:

(a) when the baby starts to lift his head: a cervical flexure with a convexity forwards develops,

(b) when the baby starts to walk: a lumbosacral flexure with a convexity forwards develops.

The *legs* and *pelvis* are small at birth. They develop with the transition to the upright posture and to walking. The pelvis shows a tilt forwards with the development of the lumbosacral convexity of the spinal column. The foot becomes everted so that the whole of it and not only the lateral border can be placed on the ground. With standing and walking the arches of the feet develop.

PUBERTY

At puberty the reproductive system matures and a number of skeletal and other changes take place. Puberty occurs in girls about the age of 10–12 years and in boys about 12–14 years. Adolescence is the period between puberty and reaching adult life a few years later.

There is a sudden spurt of growth at puberty. It starts earlier in girls, who are at first taller than boys of the same age, but as it is greater in boys men are taller than women. During puberty and adolescence girls grow about 4 cm and boys about 10 cm.

Puberty in girls
FSH produced
Graafian follicles grow
spurt of growth
external genitalia enlarge
uterus enlarges
endometrial cycle begins
menstruation begins
pubic and axillary hair grows
breasts enlarge
fat deposited in buttocks, thighs
   and breasts
pelvis develops female shape

Puberty in boys
ICSH produced
testosterone produced
spurt of growth
external genitalia enlarge
active sperm produced
larynx doubles in size
voice deepens
pubic, axillary and facial hair
   grows

ADOLESCENCE

Bone growth

Full bone growth is achieved during adolescence and early adult life. A bone continues to grow in thickness from the cells on the inner layer of the periosteum and in length by bone formation in the epiphyseal plate. In time the cartilage separating shaft and epiphysis ossifies. The times at which centres of ossification appear at which epiphyseal lines disappear are known and their presence or absence are used for dating bones.

Fusion of epiphyses with shafts
at the elbow: about 18 years
at the shoulder: about 20 years
at the wrist: about 20 years
at the hip: about 18 years
at the ankle: about 18 years
at the knee: about 20 years

The *bone-age* of a person is a term used to describe the development of a person's bones as shown on an X-ray as compared with an average person; e.g. a person of 15 years whose bones show the picture normally expected of a person of 10 has a chronological age of 15 and a bone-age of 10.

Height

Full height is reached at about 20 years with the fusing of the epiphysis at the lower end of the femur with the shaft. The growth of a person depends on:

Genetic factors. The height of a person will, in normal circumstances, be decided by his genes. Tall people tend to have tall children, short people to have short children.

Endocrine glands. Protein metabolism (which is the important factor in growth) is affected by pituitary growth hormone, thyroxine and androgens.

Diet. A deficient diet (especially of protein) will prevent growth.

Disease. Any serious or prolonged illness will prevent growth.

Red bone marrow

Red bone marrow is present now only in the flat bones of the skull, the bodies of the vertebrae, the upper ends of the humerus and femur, the ribs and the sternum.

ADULT LIFE

With the achievement of full height and physical development a person should be at the peak of his strength in early adult life.

In middle age there is a tendency, in affluent societies, for both sexes to put on weight as a result of eating and drinking too much and taking less exercise.

From about the age of 20 the pineal gland starts to become ossified and become visible on an X-ray as a small round midline structure. The sutures of the vault of the skull ossify from 30 years onwards. In the body of the sternum the union of the various bony centres, which began in puberty, is completed in early adult life. The xiphoid cartilage becomes united with the body of the sternum at about 40 years. The manubriosternal joint does not ossify until old age.

The *menopause* occurs in women between 40 and 50 years. Ovarian cells cease to respond to stimulation by pituitary hormones. Graafian follicles and ova are no longer produced, and the ovary becomes small and fibrosed. The uterine endometrial cycle and menstruation stop. The breasts and genitalia become small. Pubic and axillary hair become less.

OLD AGE

In old age:

(a) weight is reduced,

(b) the height may become less as a result of shrinking of the intervertebral bodies,

(c) the absorption of calcium from the intestine declines and decalcification of bone is likely,

(d) with loss of teeth the mandible becomes slimmer and its angle opens out,

(e) the costal cartilages ossify,

(f) the skin loses some of its elasticity,

(g) the ability to hear high pitched sounds diminishes,

(h) an arcus senilis (which may appear much earlier) appears as an opaque ring of cholesterol around the periphery of the cornea,

(i) the functional activity of organs declines, but at different rates, e.g. the heart declines more quickly than the lungs,

(j) neurones in the central nervous system continue to decline in numbers, but in a healthy brain not at a faster rate than before; memory shows a marked deterioration in some people,

(k) biological clocks may not function so accurately,

(l) the healing of damaged tissue is slow because cells and antibodies are not formed as quickly as before.

Other Books to Study

Physiology

Lippold O.C.J. & Winton F.R. (1979) *Human Physiology* 7th edition. Churchill Livingstone, Edinburgh.

Kee J.L. (1978) *Fluids and Electrolytes with Clinical Applications*. John Wiley, Chichester.

Kelman G.R. (1975) *Physiology, a Clinical Approach* 2nd edition. Churchill Livingstone, Edinburgh.

Beck M.E. (1980) *Nutrition and Dietetics for Nurses* 5th edition. Churchill Livingstone, Edinburgh.

Anatomy

Last R.J. (1977) *Anatomy* 6th edition. Churchill Livingstone, Edinburgh.

Matzke H.A. & Floyd M.F. (1979) *Synopsis of Neuroanatomy* 3rd edition. Oxford University Press.

Embryology

Beck F., Moffat D.B. & Lloyd J.B. (1973) *Human Embryology and Genetics*. Blackwell Scientific Publications, Oxford.

Haines, R.W. & Mohiuddin A. (1972) *Handbook of Human Embryology* 5th edition. Churchill Livingstone, edinburgh.

Immunology

Roitt, I.M. (1980) *Essential Immunology* 4th edition. Blackwell Scientific Publications, Oxford.

Index